Environmental Science and Engineering

Environmental Science

Series Editors

The protection of our environment is one of the most important challenges facing today's society. At the focus of efforts to solve environmental problems are strategies to determine the actual damage, to manage problems in a viable manner, and to provide technical protection. Similar to the companion subseries Environmental Engineering, Environmental Science reports the newest results of research. The subjects covered include: air pollution; water and soil pollution; renaturation of rivers; lakes and wet areas; biological ecological; and geochemical evaluation of larger regions undergoing rehabilitation; avoidance of environmental damage. The newest research results are presented in concise presentations written in easy to understand language, ready to be put into practice.

More information about this subseries at http://www.springer.com/series/3234

Sujit Mandal · Subrata Mondal

Geoinformatics and Modelling of Landslide Susceptibility and Risk

An RS & GIS-based Model Building Approach in the Eastern Himalaya

Springer

Sujit Mandal
Department of Geography
Diamond Harbour Women's University
Diamond Harbour, West Bengal, India

Subrata Mondal
Bajitpur High School
Gangarampur, West Bengal, India

ISSN 1863-5520 ISSN 1863-5539 (electronic)
Environmental Science and Engineering
ISSN 1431-6250 ISSN 2661-8222 (electronic)
Environmental Science
ISBN 978-3-030-10494-8 ISBN 978-3-030-10495-5 (eBook)
https://doi.org/10.1007/978-3-030-10495-5

This Springer imprint is published by the registered company Springer Nature Switzerland AG
The registered company address is: Gewerbestrasse 11, 6330 Cham, Switzerland

Dedicated
To all well wishers

Preface

The present book entitled *Geoinformatics and Modelling of Landslide Susceptibility and Risk—An RS & GIS-based Model Building Approach in the Eastern Himalaya* is associated with the application of geospatial tools and techniques to perform various geomorphic and statistical models for identifying spatial distribution of landslide susceptibility and risk in the Balason River Basin of Darjeeling Himalaya, West Bengal. Landslide is one of the most destructive hazards all over the world which causes large-scale damages to human lives and properties. There are a number of evidences of destructive landslide phenomena in India which caused massive destruction. Researchers tried to study landslides and produced several methods and techniques since very long to check the destructive character of landslides. In the present day, the quantitative study of geomorphic parameters and landslide locations has become a more significant aspect in landslide studies. The role of geomorphic, hydrologic, tectonic and anthropogenic parameters in the occurrences of the landslide is widely accepted. The present work is dealt with the study of various landslide conditioning parameters, i.e. slope, aspect, curvature, elevation, geomorphology, geology, soil, lineaments, drainage network, stream power index, topographic wetness index, land use and land cover and NDVI. The application of geoinformatics helped a lot to develop all these data layers and to perform various geomorphic and statistical models with regard to landslide susceptibility and risk in the Balason River Basin. The book is comprising ten chapters, i.e. Landslide: An Overview (Chap. 1), Geomorphic, Geo-tectonic and Hydrologic Attributes and Landslide Susceptibility (Chap. 2), Slope Instability Analysis Using Morphometric Parameters: A Sub-watersheds Scale Study (Chap. 3), Geomorphic Diversity and Landslide Susceptibility: A Multi-criteria Evaluation Approach (Chap. 4), Prediction of Landslide Susceptibility Using Bivariate Models (Chap. 5), Probabilistic Approaches and Landslide Susceptibility (Chap. 6), Machine Learning Models and Spatial Distribution of Landslide Susceptibility (Chap. 7), Factor's Clustering and Identification of Suitable Factor's Group Model in Landslide Susceptibility (Chap. 8), Landslide Susceptibility and Elements at Risk: A Brief Review (Chap. 9) and Comparison Between Landslide Susceptibility Models: A Critical Review and Evaluation (Chap. 10).

The present book highlighted various approaches and methods to prepare data layers, i.e. elevation, slope, aspect, curvature, geology, geomorphology, soil, distance to lineaments, lineament density, distance to drainage, drainage density, stream power index, topographic wetness index, land use and land cover, rainfall and NDVI which were analysed in detail corresponding to the Balason River Basin of Darjeeling Himalaya. In the book, an attempt has been made to analyse 13 morphometric parameters, i.e. bifurcation ratio (Rb), length of overland flow (Lof), drainage density (Dd), stream frequency (Fs), texture ratio (Rt), drainage texture (Td), compactness coefficient (Cc), constant of channel maintenance (Ccm), shape factor/basin shape (Sf), form factor (Ff), circularity ratio (Rc), elongation ratio (Re) and relief ratio (Rr) of the Balason River Basin of Darjeeling Himalaya and to prioritize sub-watersheds (SW) and to prioritize sub-watersheds using compound ranking method in relation to slope instability. To integrate various factors several models, *Information value model* (IVM), *landslide nominal risk factor model* (LNRFM), *fuzzy logic approach* (FLA), *statistical index model* (SIM), *weight of evidence model* (WoE), *certainty factor model* (CFM), machine learning tools, i.e. support vector machine (SVM) and artificial neural network model (ANNM), AND geomorphological factor group model, lithological factor group model, hydrological factor group model, triggering factor group model, protective factor group model and anthropogenic factor group model were being performed in ARC GIS environment and finally landslide susceptibility maps were prepared with a certain level of accuracy. The study demonstrated that *certainty factor model* (CFM) and *support vector machine* (SVM) are suitable models in landslide susceptibility assessment and prediction and can be used for identifying suitable locations for planning and development of the Balason River Basin of Darjeeling Himalaya.

Diamond Harbour, West Bengal, India — Sujit Mandal
Gangarampur, West Bengal, India — Subrata Mondal

Contents

Chapter 1
Landslides: An Overview

1.1 Introduction

The dynamic interplay of disturbance and succession in ecosystems are meaningfully explained by the occurrence of landslides. It is very difficult task to restore the landslide surface area because of the presence of high degree of spatial and temporal variability in soil stability and fertility (Walker et al. 2009). The variability of landslides and its destructive character have brought attention of many research scholars in logical and scientific understanding of the concept, mechanism, vulnerability and risk of landslides. Landslides can be defined as the movement of mass of rocks, earth materials, and debris down the slope under the influence of gravity by which nature finds its way of adjusting slope stability. Varnes (1978) defined landslides as the downward and upward movement of slope forming materials consisting of rock, soil, artificial fills or a combination of these materials which occurred due to the variety of process. These movements of different materials can vary from sudden and rapid collapses to slow gradual slides. Sudden and rapid events are the very dangerous than the slow movements because of a lack of warning and the velocity and its resulting force at which material can travel down the slope. These turn out to be a serious problem when they affect human and its activities (Van Westen 1993). Globally, landslides cause thousands of deaths and injuries and damages to billions of dollars (Aleotti and Chowdhury 1999). During 1903–2007 landslides caused 57,028 people deaths and affected more than 10 million people around the world (Table 1.1). The developing countries suffer the most due to the landslides event. About 95% of the landslide disasters in developing countries have been recorded and per year 0.5% of gross national product has been lost due to the landslides (Chung and Fabbri 2005). In Indian scenario, landslides mostly occur in the rainy season. The most landslides affected areas of India are the Himalayas mountain regions, the Northeastern hill ranges, the Western Ghats, the Nilgiris, the Eastern Ghats and the Vindhya. All these are mountainous regions where landslides are the significant natural denudation process and caused by several factors like geology, relief, morphology etc. and triggered

S. Mandal and S. Mondal, *Geoinformatics and Modelling of Landslide Susceptibility and Risk*, Environmental Science and Engineering,
https://doi.org/10.1007/978-3-030-10495-5_1

Table 1.1 World scenarios of landslides statistics

Continents		No. of events	Died	Injured	Homeless	Affected	Total affected	Damage US (000'S)
Africa	Total	23	745	56	7936	13,748	21,740	NA
	Avg./event		32	2	345	598	945	NA
America	Total	145	20,684	4809	186,752	4,485,037	4,676,598	1,226,927
	Avg./event		143	33	1288	30,937	32,252	8462
Asia	Total	255	18,299	3776	3,825,311	1,647,683	5,476,770	1,534,893
	Avg./event		72	15	15,001	6462	21,748	6019
Europe	Total	72	16,758	523	8625	39,376	48,524	2,487,389
	Avg./event		23	7	120	547	674	34,547
Oceania	Total	16	542	52	18,000	2963	21,015	2466
	Avg./event		34	3	1125	185	1313	154
Total		511	57,028	9216	4,046,624	6,188,807	10,244,647	5,251,675

Source EM-DAT database (1903–2007) (OFDA/CRED 2007)

by rainfall, earthquakes, rapid stream erosion, storm waves etc. (Naranjo et al. 1994; Dai et al. 2002). The occurrences of landslides over the hilly landscape of the Darjeeling Himalaya are the most serious concern which creates constant problems and serious damages to properties and lives as well as hampering the communication line every year. However, the process has been accelerated by extensive deforestation, unsystematic and unscientific constructional activities, inadequate drainage system, urban expansion, rapid changes of land use practices etc. In this hilly region, landslides commonly occurred on cut slopes, embankments alongside roads and highways and also occurred near the high-rise apartments and residential areas (Cheng et al. 2008). So, the proper management is needed to minimize the hazardous impacts of landslides events.

1.2 The Balason River Basin of Darjeeling Himalaya

The Balason river basin of Darjeeling Himalaya is characterized by a distinct geographical environment. The evolution of landforms is largely dependent on its geological structure, lithology, relief configuration, climatic characteristics and bio-physical processes. Researchers like Tamang (2013) and Lepcha (2012) carried out studies on geo-physical set up of the Balason river basin. This basin has been experiencing active denudational processes since its origin because of greater extent of relief diversity, tectonic influences, human interferences, monsoonal rainfall and so on. Free face and rectilinear facets of slope of the landforms situated in higher altitude. The geographical profile of the Balason river basin revealed the variations in landforms characteristics and slope instability.

1.2.1 Location and Administrative Set up of the Balason River Basin

The Balason river basin falls in the Darjeeling district. It is enclosed by latitude 26° 40′ N to 27° 01′ N parallel and longitude 88° 7′ E to 88° 25′ E covering the geographical area of approximately 378.45 km^2 (Fig. 1.1). The River basin is covering parts of Rangli Rangliot, Naxalbari, Matigara, Jorebunglow Sukiapokhri, Mirik and Kurseong block of Darjeeling district in West Bengal (Table 1.2).

1.2.2 Topography

The Balason river basin is a section of Bengal's hilly region situated in Darjeeling Himalaya. The elevation of the region increased from south to north, 121 m at Matigara whereas 2416 m at Lepchajagat. The network created by different ridge and valley showed dendritic drainage pattern. Inclination of slope is different in various parts of the basin. Some areas in northern part are inclined at 30°–45° whereas some places especially in the southern part are inclined at 15°–30°. The river valley has steep ungraded channel, narrow floor, steep valley etc. in the upper portion of the basin. The gradient of the river is controlled by lithology and mass movement. The steep slopes and high altitude especially in the northern parts make the areas vulnerable to slope instability whereas the southern parts of the basin are less vulnerable due to low altitude and low slope gradient.

Table 1.2 Location and administrative information of Balason river basin

Geo-location	Area of the basin (km^2)	Name of the C.D block falls in the basin	Total number of Mouza falls in the respective C.D block within the river basin
Latitude—26° 40′ N to 27° 01′ N Longitude—88° 7′ E to 88° 25′ E	378.45	Jorebunglow Sukiapokhri block	35
		Kurseong block	36
		Matigara block	40
		Mirik block	15
		Naxalbari block	3
		Rangli rangliot block	1

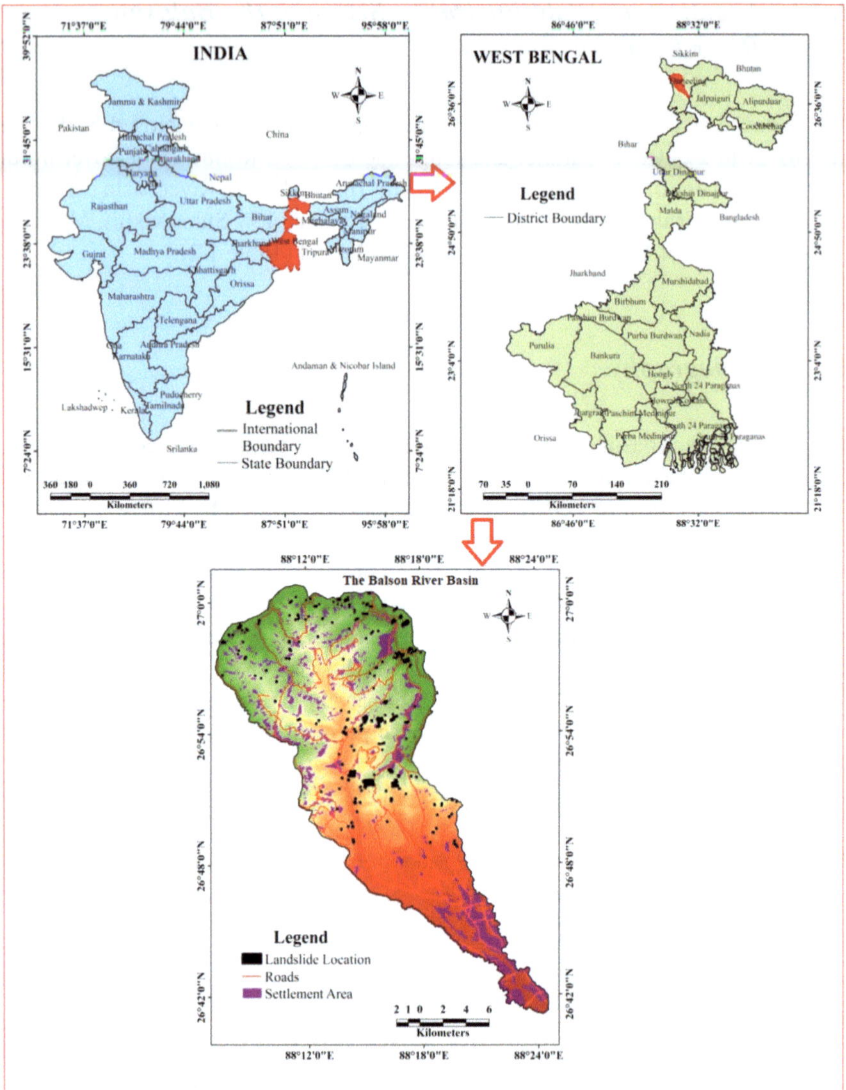

Fig. 1.1 Location map of the study area. *Source* Author Himself

1.2.3 Geology

The river basin consists of predominantly five geological structures of four different periods i.e. Pleistocene or sub-Recent to Recent, Miocene or lower Tertiary, Permian and Archaean (Table 1.3 and see Chap. 5).

It can be seen that, most of the area of this basin is covered by intensely metamorphosed rocks like gneiss, slates, phyllite and schists etc. These rocks are characterized by several fracture, joints, and faults. Also these rocks are highly weathered. Such conditions maximize the chances of landslide occurrences.

1.2.4 Geomorphology

The hydrologic characteristics of a river system determined by altitude and orientation of the river basin, ground slope of watershed, relief, soil characteristics, rock type and geologic structure (Morisawa 1968). A southern portion of the Balason river basin composed of wide open and gently sloping plain having huge amount of deposited materials associated with frequent slope failures. On the other hand the rugged and dissected hill zones are more unstable in character. However, the basin area can be divided in three distinct geomorphic zones i.e. the southern alluvial fan, the dissected foot hills, and the rugged hills (Table 1.4 and Figs. 1.2, 1.3).

1.2.5 Soil

The soil of the Balason river basin is favorable for tea plantation. The original soil of the area is drastically changed either by the process of new soil formation on the truncated top or by removal of soil by erosion (Sarkar 1990). The parent soil of this region has been changed due to its nature and it has formed more than one time (Lama 2003). In the sandy area, soils are generally siliceous and aluminous. The lower Balason basin usually formed by fine grained clay loam. In some cases the top soils are finer in nature than the subsurface soils. Based on the soil taxonomic classification, the basin is divided into 7 subgroups (Table 1.5 and Fig. 1.4). Based on the map of NBSS and LUP, five types of soil identified in the Balason river basin.

The Inceptisols soil order covers almost entire hilly tract of Balason river basin characterized by moderately shallow, well drained, gravelly loamy soils with loamy surface and moderate to severe erosion etc. which make the areas more vulnerable to slope instability.

Table 1.3 Geological periods, structures and types of rocks with some characteristics of Balason river basin

Geological periods	Geological structures	Age of beginning (million years)	Type of rocks	Some major characteristics
Pleistocene	Alluvium (older and recent)	2	Boulders beds and other sands and gravels; drift formation; younger flood plain deposits comprising sands and gravel pebble	Cover 19.5% of the basin, presence of alluvial fan
Himalayan front tectonic line				
Miocene	Nahan group; Lower Siwalik	26	Soft grayish sandstones; mudstones; shale and conglomerates along with the bands or limestone, shale and lignite	Strike WNW-ESE. with dips varying between 30° and 70°
Thrust fault (main boundary fault)				
Permian	Damuda Series (Lower Gondwana)	280	Quartzitic (hard and soft) sandstones with slaty bands; shale and slates; semi anthracitic (Graphitic coal; Lamprophyre sills and minor bands of limestone)	Coal-bearing Gondwana rocks, dip (40°–90°), strike ENE-WSW, cover 3% of the Balason basin
Thrust fault				
Archaean	Daling series	3800	Slates (greenish to grey with perfect slaty cleavage); phyllites surrounded by the pebbles of quartz; Chlorite-sericite schist with bands of gritty schist injected with gneiss (crinkled): Tourmaline and iron occur as accessories, granites pegmatite and quartz veins	The rocks are often highly metamorphosed and jointed, consistent development and monotonous lithology over a greater thickness, dip (30°–80°), cover 11% of the study area
	Darjeeling gneiss		Golden silvery mica schist; carbonaceous mica schist; granetiferous mica schist; and coarse grained gneiss	Banded appearance, texture varies from fine–grained to moderately coarse-grained rocks, dip (30°–50°), cover 59% of the basin

Source Gansser (1964), Pawde and Saha (1982), Mallet (1987)

Table 1.4 Broad geomorphic units of the Balason basin

Sl. No.	Geomorphic units	Sub types	Characteristics
1	The Southern Alluvial fan zone	a. Upper alluvial b. Lower alluvial c. River flood plain	Found between the elevation of 120–400 m, alluvial fans coalesce to form a pediment zone
2	The Dissected Foot hill zone	–	Associated with Siwalik-Damuda Sandstone and Shale, crusted and folded lithology, whole geomorphic unit has highly dissected
3	The Rugged hill zone	a. Upper hill b. Middle hill c. Lower hill	Altitude of approximately 800 m, The entire region consisted of Daling series and Darjeeling gneiss type of rocks

Table 1.5 Soil taxonomic classification of Balason river basin

Soil order	Sub order	Great groups	Sub groups	Area covered in km^2	% area to total area
Entisols	Orthents	Udorthents	Typic Udorthents	56.80	15.46
	Fluvents	Topofluvents	Typic Tropofluvents	53.26	14.50
Inceptisol	Umbrepts	Haplumbrets	Typic Haplumbrets	31.21	8.49
			Lithic Haplumbrets	25.98	7.07
			Typic Dystropepts	144.29	39.27
	Tropepts	Dystropepts	Umbic Dystropepts	38.12	10.38
			Fluventic Dystropepts	17.75	4.83

Source Lepcha (2012)

1.2.6 Climate: Rainfall and Temperature

The Balason River basin mainly experienced by four seasons namely the winter season (December–March), the summer season (April–May), the rainy season (June–September) and the post monsoon season (October–November). The seasonal duration and extent are not similar within the river basin. Monsoon is much longer in hilly region than plain which is characterized by maximum cloudy days, mist and

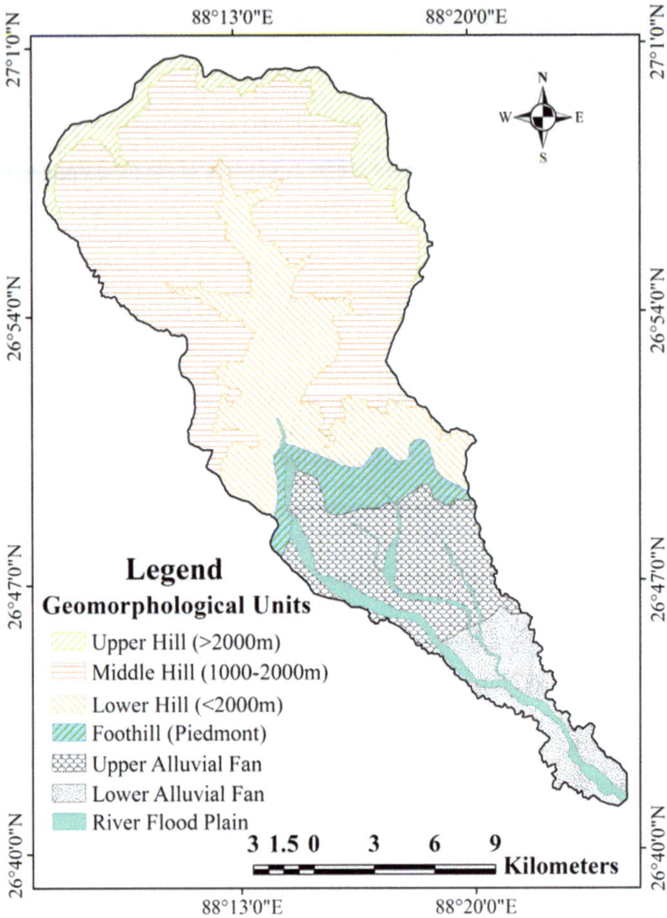

Fig. 1.2 Geomorphic units of Balason river basin. *Source* After Lama (2003)

heavy downpour etc. The cold winter season which irritates the people of the hilly region. But the climate of southern part of the basin i.e. plain is very much hot and wet in summer.

1.2.6.1 Rainfall

The rainiest month of the basin is July and the second rainiest is August. About 85% of annual rainfall is mainly occurring due to the South West Monsoon. Little amount of rainfall also served to the river basin in the month of April, May and October. Hilly regions of the basin are experienced by more rainy days (124) than plain regions (100), as more than 2.5 mm rainfall in a single day. The hilly region also

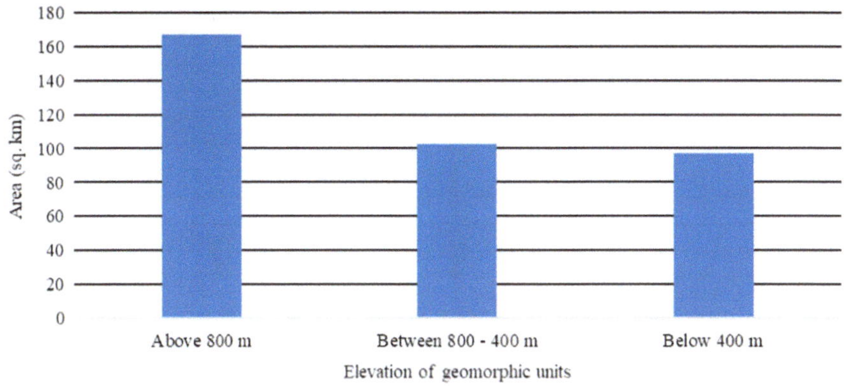

Source: After Tamang (2013)

Fig. 1.3 The areal coverage of different geomorphic units in Balason river basin. *Source* After Tamang (2013)

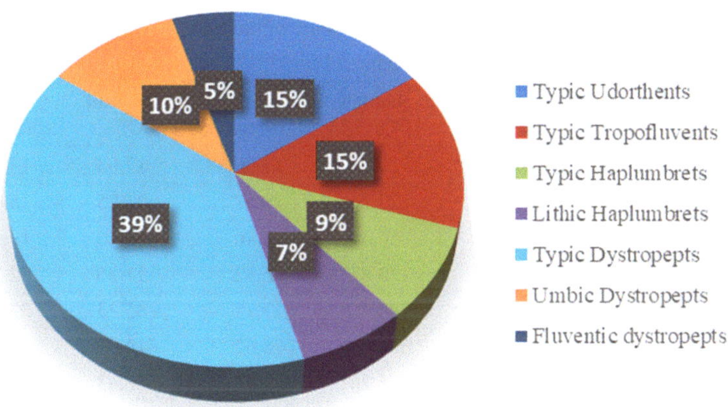

Fig. 1.4 Areal coverage of seven soil sub-groups of the Balason river basin

receives more rainfall than the plain. The whole basin receives avg. 2000–5000 mm rainfall (Table 2.5 and Fig. 1.5).

1.2.6.2 Temperature

Mean temperature of the Balason river basin varies between upper and lower segment. About 12 °C mean annual temperature is observed in the hilly region whereas recorded mean annual temperature in the plain region is 24 °C. In summer season mean temperature ranges from 27 to 29 °C in the plains and hill temperature ranges from 16 to 22 °C, whereas mean winter temperature of the plain region ranges from

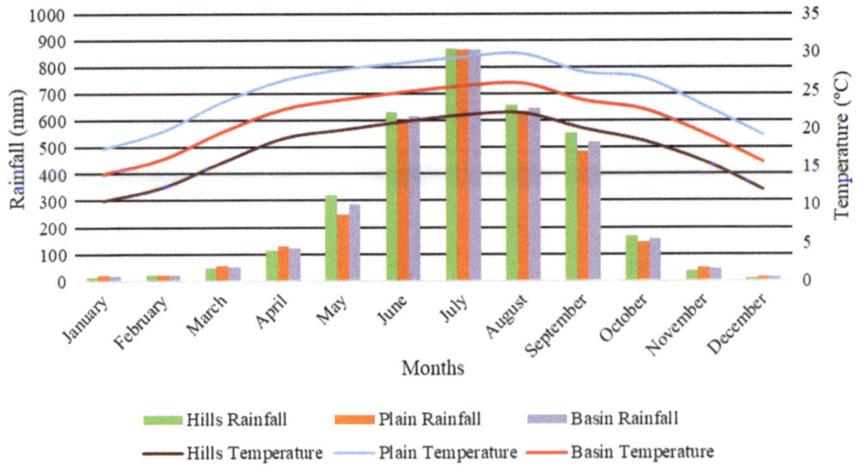

Fig. 1.5 Avg. rainfall and temperature of different months in Balason river basin (2000–2016)

Table 1.6 Average month-wise rainfall and temperature of the Balason river basin (2000–2010)

Months	Hills		Plain		Basin	
	Rainfall (°C)	Temperature (mm)	Rainfall (°C)	Temperature (mm)	Rainfall (°C)	Temperature (mm)
January	13.89	10.55	20.16	17.39	17.03	13.97
February	20.16	12.28	22.34	19.67	21.25	15.98
March	47.97	15.55	56.34	23.46	52.16	19.51
April	112.89	18.67	129.82	26.32	121.36	22.5
May	321.67	19.79	248.98	27.77	285.33	23.78
June	631.45	20.85	598.44	28.62	614.95	24.74
July	867.59	21.72	865.26	29.41	866.43	25.57
August	659.89	21.98	634.66	29.82	647.28	25.9
September	556.24	19.99	487.21	27.44	521.73	23.72
October	167.39	18.37	145.99	26.69	156.69	22.53
November	36.96	15.69	48.99	23.03	42.98	19.36
December	10.88	11.98	14.46	19.21	12.67	15.6

Source Tea garden data of Balason river basin (2000–2010)

12 to 14 °C, and 5 to 7 °C in the hilly region (Table 1.6 and Fig. 1.5). Plain region experiences with 58% humidity in the month of March and 87% in the month of June. But, in all the season, hilly area experienced more humidity than plains.

It can be seen that, the climatic characteristics of the basin are largely depend on altitude. Most of the landslides of Darjeeling Himalaya as well as Balason river basin are occurred due to the heavy rainfall especially in the hilly tracts.

Table 1.7 Drainage network of Balason river basin

Stream order	No. of stream	Stream length (km)
1st order	2744	852.89
2nd order	817	287.46
3rd order	175	145.07
4th order	37	78.3
5th order	11	58.69
6th order	3	18.19
7th order	3	36.15

1.2.7 Drainage System

The major right bank tributary of the Mahananda River, the Balason River originates from Lepchajagat (2361 m) located on the Ghum-Simana ridge. The main Balason River flows north to south-east having length of about 51.92 km and joins Mahananda River at 26° 41′ 28″ N and 88° 24′ 15″ E. The total number of stream in the Balason river basin is 3790 and total length of all segments of stream is 1476.75 km (Table 1.7). The high gradient in the upper segment of the basin increases the power of stream which leads to soil erosion and slope instability. Also the high stream density and stream frequency make the slope more vulnerable to landslides.

1.2.8 Natural Vegetation

Different kinds of forest resources as well as different species are found in the Balason river basin. This type of variation is found only very few places in India. Many scholars like Hooker (1854), Champion (1936), Banerjee, (1964), Bhujel (1996) etc. tried to classify the natural vegetation of this region. Many factors are responsible for the growth of natural vegetation of in the region where climatic and the pedological factors are most dominant. However, the illegal elimination of forest cover in the Balason river basin enhances the slope instability. In the present study area, there are some major vegetation types and sub-types (Table 1.8).

1.3 Brief Account of Landslides in Darjeeling Himalaya

Darjeeling Himalayas located in the Eastern Himalaya mountain ranges which is drained by many streams and rivers such as Tista, Rangeet, Mahannda, Jaldhaka, Balason, Mechi, Lish, Gish, Murti etc. The geological and lithological characteristics make this region high susceptible to landslide occurrences. As per the record, In Sept, 1899 Darjeeling town was badly affected by landslides. Approximately 72 people

Table 1.8 Natural vegetation of Balason river basin

Types	Sub-types	Characteristics	Location	Tree species
The Tropical and Plains vegetation (Plain to 800 m)	The Riverain forests	High temperature, Heavy rainfall, deciduous forests	Along the riverside of Balason, Rakti, Rohini and Mechi	*Meliosma pinnata, Albizia procera, Albizia lebbeck, Acacia lenticularis, Alstonia scholaris, Lagerostroemia parviflora, Oroxylum indicum* and *Globba macroclada*
	Sal (Shorea robusta) forest		Lower Siwalik, Terai and bhabar	*Terminalia alata, Aglaialawii, Duabanga grandiflora, Eugenia kurzii, Dillenia pentagynai, Tetrameles nudiflora, chelonoides, Anthocephalus chinensis* etc.
	The Dry mixed forest			*Gmelia arborea, Tetrameles nudiflora, Beilschmiedia dalzellii, Erythrina stricta, Bombax ceiba* etc.
	Wet mixed forest			*Terminalia myriocarpa, Michelia champaca, Sygzygium Formosa, Beilschmiedia roxburghiana.*
Sub-tropical forests (800–1600 m)		Deciduous forest, winter is dry and monsoon is wet	Upper Balason, Rangbang Khola, Marma Khola, Manjwa Jhora, Rakti Rohini khola	*Castanopsis indica, Schima wallichii, Gmelia arborea, Adina cordifolia, Bischofia javanica, Callicarpa arborea, Alangium chinensis*, etc. Climbers are *Bauhinia vahlii, Tinospora cordifolia, Cissampelos pareira* etc.
Temperate forest (1600–2400 m)	Temperate Deciduous forest	Widest diversity and presence of largest number of species	Kaurseong, Toong, Sonada, Darjiling, Mirik, Sukhia Pokhri, Maneyb-hangyang and Rimbick areas	*Betula alnoides, Exbucklandia populnea, Acer campbellii, Engelhardtia spicata, Lindera neesiana, Prunus napaulensis, Alnus nepalensis, Rhododendron grande, Rhododendron Eurya acuminata* etc.
	Evergreen Oak forest			*Quercus lamellose, Lithocarpus pachyphylla, Cinnamomum impressinervium* etc.

Source Tamang (2013)

were died due to this disastrous event (Table 1.4). Darjeeling, Ghum and Kurseong town were affected by Jan, 1934 landslide events. In July, 1950 about 127 people were died and roads and Siliguri–Kalimpong railways line were greatly affected due to occurrences of massive landslides. In Oct, 1968 landslides took 667 lives and damages tea garden areas and NH 31 and Hill Cart Road. Beside these, several landslide disasters occurred in the year of 1980, 1991, 1993, 2003, 2005, 2009 and 2015 in Darjeeling Himalaya (Table 1.9).

Table 1.9 Some major destructive landslide events in Darjeeling Himalaya

Year with date	Death and other losses	Affected areas
24th and 25th Sept, 1899	72 people died	Darjeeling town, Kalimpong, Kurseong, Ghum, Tindharia
15th Jan, 1934	Occurred outside the human habitation areas and so little destruction was found	Darjeeling, Ghum and Kurseong town
11th and 12th June, 1950	127 people died and thousands of people were homeless	Darjeeling town, Kalimpong, Kurseong, Happy Valley, Tindharia, Takdah
3rd and 5th Oct, 1968	Hill cart road and NH31 wiped out, heavy loss of properties and lives	Darjeeling town, Manpuri, Lebong, Kalimpong, Tista Bazar, etc.
3rd and 4th Set, 1980	More than 215 people died and 100 million properties were damaged	Rimbik, Lodhama, Darjeeling Town, Bijanbari, Lebong, Ghum, Happy Valley
16th Sept, 1991	Toy train stopped for five days and 2 people lost	Darjeeling Town, Kalimpong, Kurseong, Pulbazar, Rongtong, Tindharia etc.
13th July, 1993	15 people died	Darjeeling Town, Kalimpong, Kurseong, Pulbazar, Rongtong, Tindharia etc.
8th July, 2003	17 people died	Darjeeling Town, Kalimpong, Kurseong, Pulbazar, Rongtong, Tindharia etc.
8th July, 2005	300 homeless people and 1 died	Darjeeling Town, Kalimpong, Kurseong etc.
26th May, 2009	30 people died and loss of properties	Darjeeling, Kurseong etc.
16th Aug, 2009	7 people died and 500 houses destroyed	Nimbong, Kalimpong, Kurseong, and Mirik
2nd July, 2015	38 people killed and 23 people missing	Darjeeling, Kalimpong and Kurseong

Source Starkel and Basu (2000), Sarkar et al. (2010); http://indianexpress.com/article/india/west-bengal/landslides-claim-18-lives-in-darjeeling/

1.4 Possible Causes of Landslides in Darjeeling Himalaya

Himalaya is a very young folded mountain characterized by weak geological structure, tectonic disturbances, parallel subsidence of Himalayan fore deep of slopes etc. All these things leads to the occurrence of landslides. Darjeeling Himalaya is received by high intensity orographic rainfall every year which stimulates slope materials to move downward. Apart from this, excessive deforestation, expansion of settlements and towns especially along the roads, plying of heavy loaded vehicles, construction of new roads etc. played an important role in slope instability in Darjeeling Himalaya. So, the systematic and scientific management plans are required for the management of natural resources as well as properties and lives.

1.5 Types of Landslides

Landslide can be classified on the basis of several criteria i.e. rate of movement of materials, shape of the slope surface, the materials present in the landslips and so on. Saverenski (1937) classified landslides into three categories on the basis of shape of landslide surface i.e. a sequent landslides (develop in homogeneous cohesive soils along curved and cylindrical surfaces), consequent landslides (move along the bedding plane, joints and fractures) and insequent landslides (large and move transverse to bedding plane). On the basis of type of movement Carson and Kirkby (1971) divided into three categories such as slide, flow and heave. Slides take place along clear cut shear planes and divided into two groups i.e. translational and rotational slides. Flow is one of most destructive natural process caused due to the gravitational force. Varnes (1978) categorized flow into 4 types such as avalanches, debris flows, earth flows or mud flows. Heave can be classified as soil and frost creep and caused due to the alternative drying and wetting process, contraction and cooling process and burrowing activities of animals. Varnes (1978) classified landslides into five major groups (Table 1.10).

1.6 Mechanism and Causes of Landslides

The hill slope stability can be expressed in terms of the relationship between the forces tending to drive the hill slope materials and the forces tending to resist driving stress (Table 1.11). When driving force exceeds the resisting force, the slope materials will start to move down the slope under the influence of gravity. This relationship can be expressed in terms of safety factor (FS).

$$\mathrm{FS} = \frac{\mathrm{SS}}{\tilde{\iota}}$$

Table 1.10 Five major types of landslides

Type of movement of slides	Slides materials		
	Bedrock	Engineering soils	
		Predominantly coarse	Predominantly fine
Falls	Rock fall	Debris fall	Earth fall
Topples	Rock topple	Debris topple	Earth topple
Slides (rotational and translational)	Rock slide	Debris slide	Earth slide
Lateral spreads	Rock spread	Debris spread	Earth spread
Flows	Rock creep	Debris creep	Earth creep

Source Varnes (1978)

Table 1.11 Factors of increase of shear stress and decrease of shear strength

Factors which increase the shear stress	Factors which decrease the shear strength
Removal of lateral support by rivers, glaciers, sea wave, fault, previous rock fall or slide	Disintegration and decomposition of rocks due to weathering, hydration process etc.
Removal of underlying support by rivers, sea wave etc.	Saturation and weakening of weathered materials, alter in pore-water pressure
Slope loading due to the weight of vegetation, water and debris	Changes of structure due to the remoulding of sands, creation of fissures in shales and clays etc.
Lateral pressure due to the water present in the cracks, freezing in cracks, swelling and pressure release etc.	Organic impact in respect of burrowing of animals and dead organ of tree roots
Transient stress due to the earthquake	

Source Varnes (1978)

where ss is the total shear strength along a specific shear plane and $\tilde{\iota}$ is the total amount of shear stress developed along this plane.

The causes of landslides are usually related to slope instabilities. It is necessary to identify one or more causes and triggering factor (Table 1.12). Sometimes the terms causes and trigger create confusion. Causes may be considered as factors that make slope becoming unstable or vulnerable. On the other hand, trigger is the single event that finally initiates the movement.

Table 1.12 Landslide causes and triggering mechanisms

1. *Ground conditions*	
1	Plastic weak material, Sensitive material, Collapsible material, Weathered material, Sheared material and Jointed or fissured material
2	Adversely oriented mass discontinuities (including bedding, schistosity, cleavage) and structural discontinuities (including faults, unconformities, flexural shears, sedimentary contacts)
3	Contrast in permeability and its effects on ground water contrast in stiffness (stiff, dense material over plastic material)
2. *Geomorphological causes*	
1	Tectonic and volcanic uplift
2	Glacial rebound
3	Fluvial, glacial and wave erosion of the slope toe
4	Erosion of the lateral margins and Subterranean erosion (solution, piping)
5	Deposition loading of the slope or its crest
6	Vegetation removal (by erosion, forest fire, drought)
3. *Physical causes-triggers*	
1	Intense, short period rainfall and Prolonged high precipitation
2	Rapid melt of deep snow
3	Rapid drawdown following floods, high tides or breaching of natural dams
4	Earthquake and Volcanic eruption
5	Breaching of crater lakes and Shrink and swell weathering of expansive soils
6	Thawing of permafrost, Freeze and thaw weathering
7	Freeze-and-thaw weathering and Shrink-and-swell weathering
4. *Man-made causes*	
1	Excavation of the slope or its toe and Loading of the slope or its crest
2	Drawdown (of reservoirs) and Irrigation
3	Defective maintenance of drainage systems and Water leakage from services (water supplies, sewers, storm water drains)
4	Vegetation removal (deforestation) and Mining and quarrying (open pits or underground galleries) and Creation of dumps of very loose waste
5	Artificial vibration (including traffic, pile driving, heavy machinery)

Source Highland and Bobrowsky (2008)

1.7 Techniques and Methods of Landslide Susceptibility Assessment and Prediction: A Brief Review

When the slope becomes unstable, material is transported down the slope until a stable slope condition is attained. Slope instability is closely associated with landslide susceptibility. If slope instability is high, the chance of landslide occurrences is also high. The term landslide susceptibility refers to the possibility or chance or likelihood of potential landslides in a given area based on the local geo-environmental conditions (Brabb 1984; Grozavu et al. 2013). Magliulo et al. (2008) considered lithology, structure, slope material, vegetation cover, climate, and land use parameters for landslide susceptibility mapping. Topography factors like aspect, slope angle, plan curvature and profile curvature were used by Lee et al. (2002) for susceptibility analysis. Linear features such as faults, trails and drainage networks were used by Rowbotham and Dudycha (1998). Apart from the natural factors, human activities such as road construction, slope-cutting, urbanization and other anthropogenic activities played an important role in landslide occurrence (Sidle and Ochiai 2006). Some researchers used precipitation as a triggering factor (Larsen and Simon 1993; Stern 1995). Besides precipitation, earthquake of different intensity especially in the seismic prone areas was considered triggering factor by Garwood et al. (1979), Restrepo and Alvarez (2006).

Generally landslide susceptibility map is prepared by integrating the landslides causative data layers using various scientific models. Many statistical methods of preparing landslide susceptibility and hazard maps have been developed in the last few decades which were grouped into five broad categories, namely geomorphological mapping, landslide inventory analysis, heuristic or index-based methods, process based approaches, and statistically-based models (van Westen et al. 2003; Guzzetti et al. 2005; Lee et al. 2013). Geomorphological mapping depends on the ability of an experts investigator to analysis and map the actual and potential slope failure areas. Analysis of landslide inventories considers the known distribution of past and present landslides to delineate future landslide occurrences. In the heuristic approach, subjective knowledge is required to rank and weight the known instability factors based on their assumed or expected importance in causing landslides (Hansen et al. 1995). Process-based methods rely upon simplified which analyse the instability conditions using simple limit equilibrium models (Anagnostopoulos et al. 2015; Alvioli et al. 2016). Lastly, the statistical models were developed considering the relationship between past landslide events and the prevailing geo-environmental conditions such as bivariate models like evidential belief function (Ghosh and Carranza 2010; Tien Bui et al. 2012; Althuwaynee et al. 2012, 2014; Lee et al. 2013; Pradhan et al. 2014), frequency ratio (FR) (Yilmaz and keskin 2009; Pradhan and Lee 2010; Yalcin et al. 2011), likelihood ratio (LR) (Lee 2004; Sharma et al. 2013), information value (Info Val) method (van Westen 1997; Xu 2013; Chen et al. 2014), landslide normal risk factor (LNRF) (Gupta and Joshi 1990; Torkashvand et al. 2014), weights of Evidence (WoE) (Pradhan et al. 2010; Mohammady et al. 2012; Pourghasemi et al. 2012), fuzzy Logic (FL) (Suh et al. 2011; Guettouche 2013, Sharma et al. 2013), and statistical index (SI) models (Pourghasemi et al. 2012; Tay et al. 2014; Bourenane et al. 2015), multivariate models like logistic regression (LR) (Nandi and Shakoor 2010; Akgun et al. 2012; Ghosh et al. 2011), and discriminant analysis (DA) models (Fisher 1936; Davis 2002; Venables and Ripley 2002), pseudo quantitative approach like analytic hierarchy process (AHP) (Yalcin 2008; Ma et al. 2013; Daneshvar 2014), probabilistic approach like bayesian probability (BP) (Lee et al. 2002; Sujatha et al. 2014) and certainty factor (CF) approach (Devkota et al. 2013; Pradhan and Kim 2014; Wang et al. 2015) were frequently used by various researchers in different parts of the world which yielded reasonable landslide prediction accuracy. Recently, random forest (RF) algorithms (Kausar and Majid 2016), bayes-based algorithms such as Naïve bayes (Pham et al. 2017) and Bayes Networks (Pham et al. 2016), maximum entropy (MaxEnt) (Felicísimo et al. 2012), support vector machine (SVM) (Peng et al. 2014; Hong et al. 2015), artificial neural network (ANN) (Park et al. 2013; Tsangaratos and Benardos 2014) and decision tree method (Pradhan 2013; Naghibi et al. 2016) were used to prepare LANDSLIDE SUSCEPTIBILITY map. From 1983 to 2016, the most common statistical methods for landslide susceptibility mapping were logistic regression, neural network analysis, data-overlay, index-based and weight of evidence analyses, with an increasing preference towards machine learning methods (Reichenbach et al. 2017). In addition, nowadays researcher used LIDAR (Light Detection and Ranging) dataset instead of SRTM DEM and ASTER DEM data for the preparation of LS maps incorporating different landslide susceptibility techniques (Gutierrez et al. 2001; Slob and Hack 2004). In India, several statistical techniques were already employed to find out different landslide susceptibility zones (Basu and Pal 2017a, b; Mondal and Mandal 2017a, b; Mandal and Mandal 2018). However, there is no general way to delineate LS zones (Duman et al. 2006; Ma et al. 2013). Every method has a certain types of advantages and disadvantages. So, the landslide susceptibility methods need to be developed for authenticated landslide susceptibility maps (Ma et al. 2013). Some researchers already prepared landslide susceptibility map on small segments of Darjeeling Himalaya using statistical techniques considering several geo-environmental factors (Sarkar and Kanungo 2004; Ghosh et al. 2011; Mandal and Maiti 2014; Mondal and Mandal 2017a, b; Basu and Pal 2017a, b) and suggested some mitigation measures (Bhandari 2007; Chauhan et al. 2010; Mandal and Maiti 2014).

1.8 Relevance of Landslide Studies

In this chapter, it is already mentioned that, the occurrence of landslides in Darjeeling Himalaya at various time frame caused serious damages to properties and took many people lives. Several literatures on landslides of Darjeeling Himalaya (Nautiiyal 1951, 1966; Dutta 1950; Basu 1970; Chatterjee 1983; Basu and Sarkar 1987; Sengupta 1995; Kanungo et al. 2006; Mandal and Maiti 2014) were already stated about the enormity of landslides damages done. The future of the Darjeeling Himalaya does not look very bright. However, the Balason river basin falls in this region and is no exception to landslides phenomenon. Several landslides of various sizes occurred in this region (Fig. 1.6 and Table 1.13). Various factors like geological, geomorphological, hydrological and anthropogenic activities etc. are responsible for slope instability in this region. These factors make the areas more vulnerable. However, to mitigate or reduce the occurrences of landslide, landslide susceptibility mapping is an important step because it depicts stable and unstable area of landslide occurrences. The landslide susceptibility map can be used for developmental activities, better land use planning and management of natural resources in this area.

Fig. 1.6 Some landslides in the Balason river basin **a** near ambootia tea garden (Landslides ID 132) **b** near Dhudia (Landslides ID 133) **c** near Rishi road (Landslides ID 274) **d** near Saurinibusty (Landslides ID 273) **e** near Dhudia (Landslides ID 144) **f** near Kharia busty (Landslides ID 95)

Table 1.13 Area and location of the landslides in the Balason river basin (2000–2016)

Landslides ID	Area of the slides (m^2)	X coordinate of centroid of the slides area	Y coordinate of centroid of the slides area	Landslides Id	Area of the slides (m^2)	X coordinate of centroid of the slides area	Y coordinate of centroid of the slides area
1	707.199	88.2957	26.866	149	1062.05	88.2308	26.8763
2	1241.36	88.3238	26.8357	150	9596.98	88.2323	26.9005
3	31.7915	88.328	26.8172	151	50.3166	88.1552	26.9443
4	39.2974	88.3282	26.8173	152	1064.16	88.2108	26.9096
5	74.7919	88.3277	26.8153	153	1842.76	88.1723	26.9248
6	202.192	88.3303	26.8127	154	382.255	88.1405	26.9607
7	443.925	88.329	26.8144	155	395.708	88.1413	26.9636
8	182.456	88.1269	26.9448	156	244.159	88.1268	26.9694
9	76.8791	88.2955	26.9618	157	656.173	88.1278	26.9708
10	135.181	88.2954	26.9617	158	2051.94	88.129	26.9697
11	18.0313	88.2936	26.9614	159	826.188	88.1328	26.9755
12	4.79447	88.2935	26.9614	160	1431.73	88.1648	26.9482
13	115.793	88.3003	26.9606	161	2284.52	88.1736	26.9398
14	17.4485	88.2958	26.9627	162	244.237	88.1939	26.911
15	6.26611	88.2958	26.9624	163	473.227	88.2084	26.9073
16	61.9584	88.2918	26.9602	164	1118.3	88.1849	26.9571
17	21.4497	88.2922	26.9603	165	1343.13	88.1822	26.9573
18	20.2991	88.2894	26.9611	166	58.4183	88.1584	26.9833
19	28.2176	88.2895	26.9629	167	75.5655	88.159	26.9858
20	71.8128	88.2903	26.9623	168	114.422	88.1588	26.9855
21	23.3692	88.2905	26.9627	169	1045.16	88.1908	26.9865
22	15.4175	88.2908	26.963	170	72.5905	88.1933	26.9869
23	35.9775	88.291	26.9631	171	162.601	88.1935	26.9898
24	21.1504	88.2911	26.9633	172	1738.58	88.1929	26.99
25	92.1921	88.2908	26.9639	173	167.346	88.1937	26.988
26	32.9415	88.2913	26.9636	174	467.586	88.1856	26.9898
27	53.7706	88.2914	26.9641	175	530.686	88.1938	26.9698
28	17.3334	88.2918	26.9632	176	151.251	88.1858	26.9994
29	38.7774	88.2922	26.9635	177	108.052	88.187	27.0002
30	51.2226	88.2929	26.9635	178	158.193	88.1876	27.0007
31	68.9653	88.2852	26.9596	179	694.12	88.209	26.9963
32	379.161	88.2814	26.9609	180	509.409	88.2095	26.9967
33	1699.4	88.2809	26.9601	181	79.7281	88.2076	26.9966
34	2346.11	88.28	26.9597	182	51.9661	88.2119	26.9981
35	358.216	88.2803	26.9581	183	154.707	88.2041	27.0023
36	282.583	88.2857	26.9548	184	180.331	88.2003	27.0012
37	39.415	88.2862	26.9543	185	237.411	88.1997	26.9865
38	39.1453	88.2856	26.9538	186	192.888	88.2024	26.9888
39	22.5493	88.286	26.9536	187	1740.5	88.2106	26.9734
40	37.5993	88.2872	26.9537	188	337.62	88.243	26.9935
41	35.2606	88.288	26.954	189	383.982	88.2563	27.0058
42	48.6892	88.288	26.9544	190	31.9148	88.2553	27.006

(continued)

Table 1.13 (continued)

Landslides ID	Area of the slides (m^2)	X coordinate of centroid of the slides area	Y coordinate of centroid of the slides area	Landslides Id	Area of the slides (m^2)	X coordinate of centroid of the slides area	Y coordinate of centroid of the slides area
43	166.631	88.2877	26.9547	191	233.374	88.2374	27.0011
44	42.7794	88.2871	26.9549	192	248.928	88.2373	27.0009
45	27.9714	88.2867	26.9549	193	65.6938	88.2373	27.0031
46	22.5377	88.2868	26.955	194	1664.16	88.2354	27.0014
47	41.6531	88.287	26.9551	195	64.8215	88.2367	27.0014
48	77.0869	88.2876	26.9552	196	2662.13	88.2354	27.0007
49	7.44129	88.2882	26.9551	197	256.755	88.2352	27.0002
50	48.5958	88.2885	26.9552	198	910.659	88.2346	27.0004
51	76.4866	88.2883	26.9555	199	94.6291	88.2305	27.0015
52	23.7807	88.2881	26.9558	200	166.369	88.2298	27.001
53	179.457	88.2893	26.956	201	28.7871	88.2286	27.0013
54	41.638	88.2896	26.956	202	13.5587	88.2304	27.0019
55	42.7744	88.2895	26.9563	203	360.114	88.2283	27.0012
56	25.8678	88.2889	26.9542	204	73.1794	88.2286	27.001
57	80.355	88.2899	26.9543	205	176.405	88.2246	26.9997
58	11.6655	88.2898	26.9544	206	57.8003	88.2282	26.9979
59	13.3572	88.29	26.9543	207	29.0432	88.2317	26.9969
60	52.2182	88.2881	26.9541	208	645.099	88.2233	26.9841
61	17.4687	88.2869	26.9531	209	1045.4	88.2148	26.9871
62	75.9915	88.2793	26.9568	210	106.941	88.254	26.9982
63	201.801	88.2791	26.9567	211	217.025	88.258	26.9959
64	434.811	88.2768	26.957	212	299.998	88.2566	26.996
65	5056.89	88.2778	26.9845	213	32.9966	88.2548	26.9982
66	653.957	88.2758	26.9844	214	9.42264	88.2545	26.9977
67	133.234	88.2764	26.9853	215	45.1092	88.2539	26.9973
68	78.322	88.2765	26.9855	216	6.65395	88.2533	26.997
69	71.0966	88.2772	26.986	217	207.835	88.2505	26.9906
70	92,284.2	88.2872	26.9825	218	67.6416	88.2512	26.9901
71	7635.69	88.2885	26.9846	219	61.1047	88.253	26.9904
72	23,530.9	88.283	26.9878	220	300.908	88.2541	26.9903
73	1342.71	88.2775	26.9928	221	30.4599	88.243	26.9917
74	2953.5	88.2798	26.9944	222	133.583	88.2508	26.9986
75	20,926.3	88.2692	26.9967	223	465.7	88.2459	26.9867
76	649.424	88.2385	26.9974	224	36.006	88.2444	26.9866
77	68.4146	88.2384	26.9956	225	11.4642	88.2444	26.9865
78	982.645	88.2692	26.8557	226	37.306	88.2438	26.9857
79	315.397	88.287	26.8951	227	319.488	88.2328	26.9563
80	7041.19	88.2786	26.9122	228	602.203	88.253	26.9719
81	10,350.4	88.2876	26.9076	229	209.096	88.2575	26.9754
82	241.655	88.1912	27.0038	230	174.979	88.2584	26.9764
83	1791.87	88.1862	26.9635	231	675.168	88.2589	26.9768
84	1133.26	88.1853	26.9642	232	253.721	88.2636	26.9773

(continued)

Table 1.13 (continued)

Landslides ID	Area of the slides (m^2)	X coordinate of centroid of the slides area	Y coordinate of centroid of the slides area	Landslides Id	Area of the slides (m^2)	X coordinate of centroid of the slides area	Y coordinate of centroid of the slides area
85	7687.55	88.1851	26.9697	233	172.194	88.2756	26.9829
86	35,391.3	88.2693	26.9233	234	41.7062	88.2757	26.9832
87	8668.89	88.1469	26.9579	235	377.121	88.2381	26.9792
88	96.2151	88.2784	26.9603	236	13.6503	88.2592	26.934
89	153.184	88.2796	26.959	237	103.161	88.26	26.9435
90	327,760	88.254	26.862	238	83.3302	88.256	26.9453
91	86,276.3	88.2515	26.9106	239	39.2229	88.2623	26.9495
92	178,297	88.2394	26.869	240	11.7083	88.2612	26.9495
93	51,441.1	88.2733	26.8709	241	292.799	88.267	26.9538
94	34,320.9	88.2753	26.9101	242	75.0716	88.2831	26.9577
95	26,807.8	88.2629	26.9093	243	92.8112	88.284	26.9581
96	50,497.7	88.265	26.9098	244	141.598	88.2848	26.9585
97	1356.49	88.1428	26.9831	245	44.7715	88.2842	26.9587
98	302.838	88.2356	26.8291	246	65.5269	88.2836	26.9595
99	4286.21	88.2783	26.8624	247	30.6295	88.2834	26.9586
100	67.006	88.2906	26.9501	248	19.372	88.2831	26.9592
101	207.329	88.2915	26.9515	249	153.508	88.2832	26.9595
102	3608.52	88.2776	26.8639	250	47.3037	88.2833	26.9596
103	594.788	88.2946	26.9197	251	23.1429	88.2912	26.963
104	8094.87	88.28	26.8669	252	59.9813	88.2839	26.9556
105	293.191	88.2766	26.8611	253	156.932	88.2373	26.916
106	167.62	88.1591	26.9864	254	2103.85	88.243	26.9009
107	205.681	88.2295	26.827	255	1549.07	88.2619	26.9045
108	31.4893	88.1604	26.9424	256	63.8999	88.2639	26.9048
109	529.442	88.226	26.8423	257	251.299	88.2633	26.9045
110	6607.22	88.1428	26.9818	258	64.6839	88.2636	26.9064
111	187.824	88.142	26.9805	259	2342.79	88.2632	26.9066
112	209.821	88.1405	26.9805	260	1326.81	88.2689	26.908
113	81.1721	88.1412	26.9814	261	212.756	88.269	26.9133
114	101.34	88.145	26.9699	262	314.619	88.2756	26.9138
115	48.6915	88.1362	26.9754	263	43.8935	88.2797	26.919
116	25.7825	88.1465	26.9772	264	2120.05	88.2505	26.9079
117	165.093	88.1442	26.9789	265	242.618	88.2634	26.8916
118	93.9419	88.2303	26.917	266	252.802	88.2644	26.8913
119	717.515	88.1491	26.9788	267	262.503	88.2667	26.8917
120	458.494	88.1486	26.979	268	1719.26	88.2677	26.892
121	70.2184	88.2336	26.9036	269	43.2021	88.2662	26.887
122	548.314	88.1479	26.9797	270	27.6787	88.259	26.8683
123	1859.56	88.2423	26.85	271	6.49692	88.2282	26.8839

(continued)

Table 1.13 (continued)

Landslides ID	Area of the slides (m^2)	X coordinate of centroid of the slides area	Y coordinate of centroid of the slides area	Landslides Id	Area of the slides (m^2)	X coordinate of centroid of the slides area	Y coordinate of centroid of the slides area
124	29,211	88.2751	26.8605	272	407.037	88.2477	26.8563
125	854.153	88.2477	26.9021	273	279.718	88.2122	26.8555
126	11,015	88.2418	26.8472	274	28.7389	88.2281	26.8577
127	328.532	88.1923	26.9103	275	5350.87	88.2772	26.8624
128	252.397	88.2343	26.9044	276	30,543.9	88.2777	26.8597
129	3010.4	88.2363	26.8644	277	243.052	88.2809	26.8573
130	1247.26	88.2468	26.8589	278	260.571	88.2803	26.8575
131	789.487	88.2464	26.8472	279	212.472	88.2862	26.8534
132	14,854.8	88.2509	26.8593	280	14,374.7	88.2736	26.8491
133	1947.98	88.2402	26.8172	281	289.395	88.2698	26.8325
134	416.474	88.2414	26.8869	282	402.556	88.2435	26.8402
135	63.2228	88.2392	26.8897	283	196.266	88.3069	26.8162
136	842.363	88.2418	26.9028	284	1235.7	88.3111	26.8451
137	179.9	88.2487	26.906	285	603.006	88.291	26.8667
138	245.819	88.2485	26.903	286	571.366	88.3095	26.8239
139	1051.9	88.2476	26.9052	287	220.043	88.3099	26.8238
140	424.025	88.2481	26.9056	288	500.193	88.3123	26.8269
141	3158.76	88.236	26.9102	289	497.139	88.3259	26.8376
142	1685.25	88.2513	26.908	290	193.821	88.3303	26.8471
143	4283.61	88.2573	26.909	291	25.5029	88.3269	26.8482
144	826.283	88.2398	26.8187	292	16.4798	88.3262	26.8485
145	2661.94	88.2394	26.8289	293	142.254	88.3176	26.8458
146	440.653	88.2386	26.8318	294	54.7755	88.3173	26.8442
147	8140.19	88.2341	26.8994	295	52.0469	88.3134	26.8371
148	803.776	88.2306	26.8658				

References

Akgun A, Sezer EA, Nefeslioglu HA, Gokceoglu C, Pradhan B (2012) An easy-to-use MATLAB program (MamLand) for the assessment of landslide susceptibility using a Mamdani fuzzy algorithm. Comput Geosci 38:23–34

Aleotti P, Chowdhury R (1999) Landslide hazard assessment: summary review and new perspectives. Bull Eng Geol Environ 58:21–44

Althuwaynee OF, Pradhan B, Lee S (2012) Application of an evidential belief function model in landslide susceptibility mapping. Comput Geosci 44:120–135. https://doi.org/10.1016/j.cageo.2012.3

Althuwaynee OF, Pradhan B, Park HJ, Lee JH (2014) A novel ensemble bivariate statistical evidential belief function with knowledge-based analytical hierarchy process and multivariate statistical logistic regression for landslide susceptibility mapping. Catena 114:21–36

Alvioli M, Marchesini M, Reichenbach P, Rossi M, Ardizzone F, Fiorucci F, Guzzetti F (2016) Automatic delineation of geomorphological slope units with r.slopeunits v1.0 and their optimization for landslide susceptibility modelling. Geosci Model Dev 9:3975–3991. https://doi.org/10.5194/gmd-9-3975-2016
Anagnostopoulos GG, Fatichi S, Burlando P (2015) An advanced process based distributed model for the investigation of rainfall-induced landslides: the effect of process representation and boundary conditions. Water Resour Res 51(9):7501–7523. https://doi.org/10.1002/2015WR016909
Banerjee U (1964) A short note on the forests of Darjeeling district, West Bengal Forests Centenary Commemoration Volume. Government of West Bengal, West Bengal, pp 91–95
Basu AR (1970) A geotechnical note on investigations of certain stretches of Border roads under Swastik Project in Sikkim and Darjeeling District, West Bengal, Unpublished GSI Report, 1970, FS 1969–1970
Basu SR, Sarkar S (1987) Ecosystem vis-a vis landslide: a case study in Darjeeling Himalaya, proceed. Impact of Development on Environment, Unpublished GSI Report, FS 1981–82, 1982–83. Geog. Soc. India, vol 2, pp 45–53
Basu T, Pal S (2017a) Identification of landslide susceptibility zones in Gish River basin. Assessment and Management of Risk for Engineered Systems and Geohazards, West Bengal, India, Georisk. https://doi.org/10.1080/17499518.2017.1343482
Basu T, Pal S (2017b) Exploring landslide susceptible zones by analytic hierarchy process (AHP) for the Gish River Basin, West Bengal, India, 2017 Spat Res Inf. https://doi.org/10.1007/s41324-0170134-2
Bhandari RK (2007) The Indian landslide scenario: strategic issues and action points. Disaster & Development Vol. 1, No. 2:May
Bhujel RB (1996) Studies on the Dicotyledonous flora of Darjeeling district, unpublished Ph.D. thesis, University of North Bengal
Bourenane H, Bouhadad Y, Guettouche MS, Braham M (2015) GIS based landslide susceptibility zonation using bivariate statistical and expert approaches in the city of Constantine northeast Algeria. Bull Eng Geol Environ 74(2):337–355
Brabb EE (1984) Innovative approaches to landslide hazard and risk mapping. In: Proceedings of the fourth international symposium on landslides, Canadian Geotechnical Society, Toronto, Canada, 16–21 Sept 1984, pp 307–324
Carson MA, Kirkby MJ (1971) Hillslope form and process. Cambridge Publisher, p 475
Champion HG (1936) A preliminary survey of the forest types of India and Burma. Indian For Rec Seer. 2, Silviculture 1(1):1–286
Chatterjee B (1983) A geological approach to the landslide hazard zonation in Darjeeling Himalaya, W.B. Unpublished GSI report, FS 1981–82, 1982–83
Chauhan S, Sharma M, Arora MK (2010) Landslide susceptibility conation of the Chamoli Region, Garhwal Himalayas, using Logistic Regression Model. Landslides 7:411–423
Chen W, Li W, Hou E, Zhao Z, Deng N, Bai H, Wang D (2014) Landslide susceptibility mapping based on GIS and information value model for the Chencang District of Baoji, China, 2014. Arab J Geosci https://doi.org/10.1007/s12517-014-1369-z
Cheng TA, Lateh H, Peng KS (2008) Intelligence explanation system on landslide dissemination: a case study in Malaysia. In: Proceedings of the first world landslide forum report: implementing the 2006 Tokyo action plan on the International program on landslides (IPL), pp 330–333
Chung CJ, Fabbri AG (2005) Systematic procedures of landslide hazard mapping for risk assessment using Spatial Prediction Models. In: Glade T, Anderson MG, Crozier MJ (eds) Landslide hazard and risk. Wiley, London, pp 139–174
Dai FC, Lee CF, Ngai YY (2002) Landslide risk assessment and management: an overview. Eng Geol 64(1):65–87
Daneshvar MRM (2014) Landslide susceptibility zonation using analytical hierarchy process and GIS for the Bojnurd region, northeast of Iran. Landslides 11(6):1079–1091
Davis JC (2002) Statistics and data analysis in geology, 3rd ed. JohnWiley and Sons, 638 pp. In: ERDAS 2007, ERDAS Field Guide, 5th ed. ERDAS Inc, Atlanta, Georgia, USA, 672 pp

Devkota KC, Regmi AD, Pourghasemi HR et al (2013) Landslide susceptibility mapping using certainty factor, index of entropy and logistic regression models in GIS and their comparison at Mugling-Narayanghat road section in Nepal Himalaya. Natural Hazards 65(1):135–165. https://doi.org/10.1007/s11069-012-0347-6

Duman TY, Can T, Gokceoglu C, Nefeslioglu HA, Sonmez H (2006) Application of logistic regression for landslide susceptibility zoning of Cekmece Area, Istanbul, Turkey. Environ Geol 51:241–256

Dutta KK (1950) Landslips in Darjeeling & Neighbouring hills slopes in June. Bull Geol Surv India 15(1966):7–30

Felicísimo A, Cuartero A, Remondo J, Quirós E (2012) Mapping landslide susceptibility with logistic regression, multiple adaptive regression splines, classification and regression trees, and maximum entropy methods: a comparative study, Landslides. https://doi.org/10.1007/s10346-012-0320-1

Fisher RA (1936) The use of multiple measurements in taxonomic problems. Ann Eugenics 7:179–188

Gansser A (1964) Geology of the Himalayas. Inter Science Publishers, pp 31–97

Garwood NC, Janos DP, Brokaw N (1979) Earthquake-caused landslides: a major disturbance to tropical forests. Science 205:997–999

Ghosh S, Carranza EJM (2010) Spatial analysis of mutual fault/fracture and slope controls on rocksliding in Darjeeling Himalaya, India. Geomorphology 122(1):1–24

Ghosh S, Carranza EJM, van Westen CJ, Jetten VG, Bhattacharya DN (2011) Selecting and weighting spatial predictors for empirical modeling of landslide susceptibility in the Darjeeling Himalayas (India). Geomorphology 131(1):35–56

Grozavu A, Plescan S, Patriche CV, Margarint MC, Rosca B (2013) Landslide susceptibility assessment: GIS application to a complex mountainous environment. In: The Carpathians: Integrating nature and society towards sustainability, Environ Sci Eng, pp 31–44

Guettouche MS (2013) Modeling and risk assessment of landslides using fuzzy logic: application on the slopes of the Algerian Tell (Algeria). Arabian J Geosci 6:3163–3173

Gupta RP, Joshi BC (1990) Landslide hazard zonation using the GIS approach—a case study from the Ramganga Catchment, Himalayas. Eng Geol 28:119–131

Gutierrez R, Gibeaut J, Smyth R, Hepner T, Andrews J, Weed C, Gutelius W, Mastin M (2001) Precise airborne lidar surveying for coastal research and geohazards applications. Int Arch Photogramm Remote Sens Spat Inform Sci 34:185–194

Guzzetti P, Reichenbach M, Cardinali M, Galli F, Ardizzone F (2005) Landslide hazard assessment in the Staffora basin, Northern Italian Apennines. Geomorphology 72:272–299

Hansen A, Franks CAM, Kirk PA, Brimicombe AJ, Tung F (1995) Application of GIS to hazard assessment, with particular reference to landslidesin Hong Kong. In: Carrara A, Guzzetti F (eds) Geographical information systems in assessing natural hazards. Kluwer Academic Publisher, Dordrecht, The Netherlands, pp 135–175

Highland LM, Bobrowsky P (2008) The landslide handbook—a guide to understanding landslides: Reston, Virginia, U.S. Geol Surv Circ 1325:129

Hong H, Pradhan B, Xu C, Bui DT (2015) Spatial prediction of landslide hazard at the Yihuang area (China) using two-class kernel logistic regression, alternating decision tree and support vector machines. Catena 133:266–281

Hooker JD (1854) Himalayan journals, vol 2. John Murray, London

Kanungo DP, Arora MK, Sarkar BS, Gupta RP (2006) A comparative study of conventional, ANN black box, fuzzy and combined neural and fuzzy weighting procedures for landslide susceptibility zonation in Darjeeling Himalayas. Eng Geol 85:347–366

Kausar N, Majid A (2016) Random forest-based scheme using feature and decision levels information for multi-focus image fusion. Pattern. Anal. Applic. 19(1):221–236

Lama IL (2003) Study of the environmental geomorphology in the Balason Basin, unpublished Ph. D thesis, North Bengal University, pp 24–28

Larsen MC, Simon A (1993) A rainfall intensity-duration threshold for landslides in a humid-tropical environment. P R Geog Ann 75A
Lee S, Choi J, Min K (2002) Landslide susceptibility analysis and verification using the Bayesian probability model. Environ Geol 43:120–131
Lee S, Jaewon C, Kyungduck M (2002) Landslide susceptibility analysis and verification using the Bayesian probability model Environmental. Geology 43:120–131
Lee S (2004) Application of likelihood ratio and logistic regression models to landslide susceptibility mapping using GIS. Environ Manag 34(2):223–232
Lee S, Hwang J, Park I (2013) Application of data-driven evidential belief functions to landslide susceptibility mapping in Jinbu, Korea. Catena 100:15–30
Lepcha I (2012) Regional study of geo-environmental aspects: a study of Balason Basin, Darjeeling district, West Bengal. J Geo-Environ Observer 1(2):24–37
Ma F, Wang J, Yuan R, Zhao H, Guo J (2013) Application of analytical hierarchy process and least-squares method for landslide susceptibility assessment along the Zhong-Wu natural gas pipeline. China; Landslides 10(4):481–492
Magliulo P, Di Lisio A, Russo F, Zelano A (2008) Geomorphology and landslide susceptibility assessment using GIS and bivariate statistics, a case study in southern Italy. Nat Hazards 47:411–435
Mallet FR (1987) On the geology and mineral resources of the Darjeeling district and western Duars. Mem Geol Surv India 41(11):1–72
Mandal S, Maiti R (2014) Semi-quantitative approaches for landslide assessment and prediction. Part of the series Springer natural hazards
Mandal S, Mandal K (2018) Modeling and mapping landslide susceptibility zones using GIS based multivariate binary logistic regression (LR) model in the Rorachu river basin of eastern Sikkim Himalaya, India. Model Earth Syst Environ. https://doi.org/10.1007/s40808-018-0426-0
Mohammady M, Pourghasemi HR, Pradhan B (2012) Landslide susceptibility mapping at Golestan Province, Iran: a comparison between frequency ratio, Dempster-Shafer, and weights-of-evidence models. J Asian Earth Sci 61:221–236
Mondal S, Mandal S (2017a) RS & GIS-based landslide susceptibility mapping of the Balason River basin, Darjeeling Himalaya, using logistic regression (LR) model. Georisk: Assess Manag Risk Eng Syst Geohazards 12(1):29–44. https://doi.org/10.1080/17499518.2017.1347949
Mondal S, Mandal S (2017b) Application of frequency ratio (FR) model in spatial prediction of landslides in the Balason river basin, Darjeeling Himalaya. Spat Inf Res. https://doi.org/10.1007/s41324-017-0101-y
Morisawa M (1968) Streams: their dynamics and morphology. McGraw-Hill Book Co., New York
Naghibi SA, Pourghasemi HR, Dixon B (2016) GIS-based groundwater potential mapping using boosted regression tree, classification and regression tree, and random forest machine learning models in Iran. Environ Monit Assess 188(1):1–27
Nandi A, Shakoor A (2010) A GIS-based landslide susceptibility evaluation using bivariate and multivariate statistical analyses. Eng Geol 110(1):11–20
Naranjo JL, van Westen CJ, Soeters R (1994) Evaluating the use of training areas in bivariate statistical landslide hazard analysis—a case study in Columbia. ITC J 3
Nautiiyal SP (1966) On the stability of certain hill slopes in and around Darjeeling, W.B. Bull Geol Surv India 15:31–48
Nautiyal SP (1951) A geological report on the hill slope stability in and around Darjeeling, W.B., unpublished report of the geological survey of India
Park S, Choi C, Kim B, Kim J (2013) Landslide susceptibility mapping using frequency ratio, analytic hierarchy process, logistic regression, and artificial neural network methods at the Inje area. Korea; Environ Earth Sci 68:1443–1464
Pawde MB, Saha SS (1982) Geology of the Darjeeling Himalayan section. Geol Surv India 41(11):50–54

Peng L, Niu R, Huang B, Wu X, Zhao Y, Ye R (2014) Landslide susceptibility mapping based on rough set theory and support vector machines: a case of the Three Gorges area, China. Geomorphology 204:287–301

Pham BT, Tien Bui D, Pham HV (2016) Spatial prediction of rainfall induced landslides using Bayesian Network at Luc Yen District, Yen Bai Province (Viet Nam). In: International conference on environmental issues in mining and natural resources development (EMNR 2016). Hanoi University of mining and geology (HUMG), Viet Nam, pp 1–10

Pham BT, Bui DT, Pourghasemi HR, Indra P, Dholakia M (2017) Landslide susceptibility assessment in the Uttarakhand area (India) using GIS: a comparison study of prediction capability of naïve bayes, multilayer perceptron neural networks, and functional trees methods. Theoretical and Applied Climatology. 128:255–273

Pourghasemi RH, Pradhan B, Gokceoglu C (2012) Application of fuzzy logic and analytical hierarchy process (AHP) to landslide susceptibility mapping at Haraz watershed, Iran. Nat Hazards 63(2):965–996

Pradhan B, Lee S (2010) Landslide susceptibility assessment and factor effect analysis: backpropagation artificial neural networks and their comparison with frequency ratio and bivariate logistic regression modeling. Environ Model Softw 25:747–759

Pradhan B, Oh HJ, Buchroithner M (2010) Weights-of-evidence model applied to landslide susceptibility mapping in a tropical hilly area. Geomat Nat Haz Risk 1(3):199–223

Pradhan B (2013) A comparative study on the predictive ability of the decision tree, support vector machine and neuro-fuzzy models in landslide susceptibility mapping using GIS. Comput Geosci 51:350–365

Pradhan AMS, Kim YT (2014) Relative effect method of landslide susceptibility zonation in weathered granite soil: a case study in Deokjeok-ri Creek, South Korea. Nat Hazards 72(2):1189–1217

Reichenbach P, Rossi M, Malamud B, Mihir M, Guzzetti F (2017) A review of statistically-based landslide susceptibility models. Earth Sci Rev. https://doi.org/10.1016/j.earscirev.2018.03.001

Restrepo C, Álvarez N (2006) Landslides and their contribution to land cover in the mountain of Mexico and Central America. Biotropica 38:446–457

Rowbotham DN, Dudycha D (1998) GIS modelling of slope stability in Phewa Tal watershed, Nepal. Geomorphology 26:151–170

Sarkar S (1990) Genesis and classification of the soils of the Mahananda basin, Darjeeling Himalaya. Geog Mem 2(182):117–129

Sarkar S, Kanungo DP (2004) An integrated approach for landslide susceptibility mapping using remote sensing and GIS. Int J Photogrammetric Eng Remote Sens 70:617–625

Sarkar K, Singh TN, Verma AK (2010) A numerical simulation of landslide-prone slope in Himalayan region—a case study. Arab J Geosci 5(1):73–81. https://doi.org/10.1007/s12517-010-0148-8

Sengupta CK (1995) Detailed study of geofactors in selected hazard prone stretches along the surface communication routes in parts of Darjeeling Himalaya. Unpublished GSI report, FS 1993–94 & 1994–95

Sharma LP, Nilanchal Patel, Ghose MK, Debnath P (2013) Synergistic application of fuzzy logic and geoinformatics for landslide vulnerability zonation—a case study in Sikkim Himalayas, India. Appl Geomat 5:271–284

Slob S, Hack R (2004) 3D terrestrial laser scanning as a new field measurement and monitoring technique. In: Engineering geology for infrastructure planning in Europe: a European perspective, Lectures Notes in Earth Sciences, Springer, Berlin/Heidelberg, 104:179–189

Sidle RC, Ochiai H (2006) Landslide processes, predictions and land use, American geographical union. Water Res 18:312

Starkel L, Basu SR (2000) Rains, landslides and floods in the Darjiling Himalayas. Indian Science Academy, New Delhi, pp 1–168

Stern M (1995) Vegetative recovery on earthquake triggered landslide sites in the Ecuadorian Andes. In: Churchill SP, Balslev H, Forero E, Luteyn JL (eds) Biodiversity and conservation of neotropicalmontane forests. The New York Botanical Garden, Bronx, pp 207–220

Suh J, Choi Y, Roh TD, Lee HJ, Park HD (2011) National-scale assessment of landslide susceptibility to rank the vulnerability to failure of rock-cut slopes along expressways in Korea. Environ Earth Sci 63(3):619–632
Sujatha ER, Kumaravel P, Rajamanickam GV (2014) Bull Eng Geol Environ 73:147. https://doi.org/10.1007/s10064-013-0537-9
Tamang L (2013) Effects of boulder lifting on the fluvial characteristics of lower Balason basin in Darjeeling district west Bengal, Ph.D. thesis, University of North Bengal, Department of Geography and Applied Geography, p 19
Tay LT, Lateh H, Hossain MK, Kamil AA (2014) Landslide hazard mapping using a Poisson distribution: a case study in Penang Island, Malaysia. In Landslide science for a safer geoenvironment. Springer International Publishing, pp 521–525
Tien Bui D, Pradhan B, Lofman O, Revhaug I (2012) Landslide susceptibility assessment in Vietnam using Support vector machines, Decision tree and Naïve Bayes models. Mathematical Problems in Engineering. https://doi.org/10.1155/2012/9746382012:26
Torkashvand AM, Irani A, Sorur J (2014) The preparation of landslide map by Landslide Numerical Risk Factor (LNRF) model and Geographic Information System (GIS). Egypt J Remote Sens Space Sci 17:159–170
Tsangaratos P, Benardos A (2014) Estimating landslide susceptibility through an artificial neural network classifier. Nat Hazards 74(3):1–28
Van Westen CJ (1993) Application of geographical information systems to landslide hazard zonation, ITC-Publication No. 15, ITC, Enschede, The Netherlands, p 245
Van Westen CJ (1997) Statistical landslide hazard analysis. In: Application guide, ILWIS 2.1 for Windows. ITC, Enschede, The Netherlands, pp 73–84
Van Westen CJ, Rengers N, Soeters R (2003) Use of geomorphological information in indirect landslide susceptibility assessment. Nat Hazards 30(3):399–419
Varnes DJ (1978) Slope movement types and processes. In: Schuster RL, Krizek RJ (eds) Landslides analysis and control. National Academy Press, Washington DC, Transportation Research Board Special Report 176, pp 11–33
Venables WN, Ripley BD (2002) Modern applied statistics with S, 4th ed. Springer, Berlin, Germany, 495 pp
Walker LR, Velázquez E, Shiels AB (2009) Applying lessons from ecological succession to the restoration of landslides. Plant Soil 324:157–168
Wang Q, Li W, Chen W, Bai H (2015) GIS-based assessment of landslide susceptibility using certainty factor and index of entropy models for the Qianyang County of Baoji city, China. J Earth Syst Sci 124(7):1399–1415
Xu C (2013). Assessment of earthquake-triggered landslide susceptibility based on expert knowledge and information value methods: a case study of the 20 April 2013 Lushan, China Mw6. 6 earthquake. Disaster Advances 6(13):119–130
Yalcin A (2008) GIS-based landslide susceptibility mapping using analytical hierarchy process and bivariate statistics in Ardesen (Turkey): comparisons of results and confirmations. Catena 72:1–12
Yalcin A, Reis S, Aydinoglu AC, Yomralioglu T (2011) A GIS-based comparative study of frequency ratio, analytical hierarchy process, bivariate statistics and logistics regression methods for landslide susceptibility mapping in Trabzon, NE Turkey. Catena 85:274–287
Yilmaz I, Keskin I (2009) GIS based statistical and physical approaches to landslide susceptibility mapping (Şebinkarahisar, Turkey). Bull Eng Geol Environ 68(4):459–471

Chapter 2
Geomorphic, Geo-tectonic and Hydrologic Attributes and Landslide Susceptibility

2.1 Introduction

Landslides can be defined as the movement of mass of rock, debris or earth materials downward slope under the influence of gravitational force. Many researchers in India (Nautiiyal 1966; Sengupta 2000; Mandal and Maiti 2014) carried out demand based studies in mountainous and hilly areas and tried to find out the causative factors and consequences of landslide occurrences and recommended some mitigation practices to reduce its hazardous impact. In Darjeeling, the spatial extents of landslides are increasing day by day and causing severe damage to lives and properties. The Balason river basin is not an exception to it. The evolution of landforms is largely dependent on its geological structure, lithology, relief configuration, climatic characteristics and bio-physical processes. Researchers like Tamang (2013) and Lepcha (2012) carried out studies on geo-physical set up of Balason River basin. This basin has been experiencing active denudation processes because of high relief diversity, tectonic activities, human interferences, drainage development, and high intensity rainfall. Free face and rectilinear facets of slope of the landforms situated in higher altitude where cliffs, spurs and tea garden areas are more very much prone to landslide phenomena. However, the study made an attempt to understand the role of various geomorphic, geo-tectonic and hydrologic attributes i.e. elevation, slope, aspect, curvature, geomorphology, geology, lineament density, distance to lineament, land use/land cover (LULC), soil, normalized differential vegetation index (NDVI), drainage density, distance to drainage, stream power index (SPI), topographic wetted index (TWI) and rainfall in landslide occurrences in the Balason river basin of Darjeeling Himalaya. The probability of landslides of each class of the landslide causative factors were estimated with the help of frequency ratio (FR) which is derived by landslide affected pixels and landslide non-affected pixels.

S. Mandal and S. Mondal, *Geoinformatics and Modelling of Landslide Susceptibility and Risk*, Environmental Science and Engineering,
https://doi.org/10.1007/978-3-030-10495-5_2

2.2 Materials and Methods

In the present work, sixteen factors i.e. elevation, slope, aspect, curvature, geomorphology, geology, lineament density, distance to lineament, land use/land cover (LULC), soil, normalized differential vegetation index (NDVI), drainage density, distance to drainage, stream power index (SPI), topographic wetted index (TWI) and rainfall have been used to assess the role in landslide susceptibility. Basically landslide susceptibility refers to the possibility or chance or likelihood of potential landslides in a given area of interest based on the local terrain conditions (Grozavu et al. 2013).

Altitude, aspect, slope and curvature data layers were directly derived from SRTM DEM (30 m spatial resolution) in Arc GIS 10.1 environment which downloaded using https://earthexplorer.usgs.gov link. Geomorphological map was digitized from Unpublished Ph.D. Thesis of Lama, 2003 using Arc GIS environment. The geological map was collected from GSI (Geological Survey of India, Kolkata). The lineament was extracted from Landsat TM image (Feb, 2009) using Geomatica 2016 software. To derive distance to lineament map, multiple buffer tool in Arc Map 10.1 was used. Four major land use and land cover classes i.e. agricultural plantation, water bodies, settlements and forest have been identified from Landsat TM image acquired on Feb, 2009 and then verified using Google earth. Soil map was collected from NBSS (National Bureau of Soil Science and Land Use Planning, Kolkata) and digitized in Arc GIS 10.1 platform. NDVI map was prepared from Landsat TM (Feb, 2009) using Erdas Imagine 9.2 software. The NDVI was calculated using the formula NDVI = (IR − R)/ (IR + R), where IR is the infrared portion and R is the Red portion of the electromagnetic spectrum. Drainage network was digitized from Survey of India (SOI), Kolkata. Map no. 78a/4, 78a/8, 78b/1, 78b/5 and 78b/6 and it was consulted with and google earth image (2015). Inverse distance weighted (IDW) method was incorporated to prepare drainage density map. Distance to drainage map was prepared using multiple buffer tool in Arc GIS 10.1 software. SPI map and TWI map was prepared in Arc Map 10.1 environment using following equation:

$$\mathrm{SPI} = \mathrm{Ln}\,\{(A_s + 0.001) * (\tan\beta \text{ in percent\_rise/100}) + 0.001)\} \quad (2.1)$$

$$\mathrm{TWI} = \mathrm{Ln}\{(A_s + 0.001)/(\tan\beta \text{ in percent\_rise/100}) + 0.001)\} \quad (2.2)$$

where Ln is the natural log, A_s is the flow accumulation and $\tan\beta$ is the slope. To obtain SPI and TWI maps of the study area, the sinks of the SRTM DEM image were removed and then the slope map was prepared using raster surface tool. Then, flow accumulation was prepared from the flow direction image using the depression less DEM. Rainfall data was downloaded from http://www.worldclim.org link in Continuous raster data format and imported in Arc GIS environment.. The database of landslide distribution map was generated from Google earth historical imageries (2000–2016), Landsat 7 ETM+ imageries (2000–2005), Landsat 4-5 TM imageries (2006–2007, 2009–2011), LISS III imageries (2008, 2012–2013), Landsat 8 OLI

(2014–2015), High resolution Sentinel-2 imagery (2016), Arc Map high resolution world imagery and intensive field survey data (2015–2016). Total 295 landslide polygons were mapped in a mapped area of 378.45 km^2. The average area of the landslides was 3762.90 m^2.

To establish the relation between each factor and landslide occurrences, the frequency ratio (FR) value was used (Karim et al. 2011).

$$\mathrm{Fr_i} = \frac{\{\mathrm{N\ pix\ (Si)/N\ pix\ (Ni)}\} \times 100}{\left\{\sum \mathrm{N\ pix\ (Si)}/\sum \mathrm{N\ pix\ (Ni)}\right\} \times 100} \tag{2.3}$$

where, N pix (Si) is the number of pixels containing landslide in each class (i), N pix (Ni) is the total number of pixels having class (i) in the whole basin, or simply the percentage of landslide pixels to total pixels in a class, $\sum$N pix (Si) total number of pixels in each class and $\sum$N pix (Ni) total number of pixels in the whole basin or simply percentage of pixels to total pixels in class. To calculate the number of landslide pixels of each class of the respective factor, zonal statistics as a table under spatial analyst tool in Arc GIS 10.1 environment was used. In can be mentioned that, higher the frequency ratio value and stronger the relationship between landslide occurrence and the given conditioning factor and greater the probability of landslide phenomena.

2.2.1 Elevation Character and Landslide Potentiality

The elevation of the Balason river basin was classified into 10 categories where elevation ranges from 106 to 2603 m (Fig. 2.1a). Extreme north-western part and north-eastern part of the basin was dominated by higher elevation whereas southernmost section was characterized by low elevation. However, there is no direct relationship can be made between elevation and landslide occurrences. Several researchers showed that, landslides have the more tendencies to occur at the higher El areas, so the high El areas have the greater landslide susceptibility (Devkota et al. 2013; Umar et al. 2014). Moderate elevation zones were found in the central portion of the basin. The highest and lowest frequency ratio value was found in the elevation class of 1065–1257 m (2.79) and 106–267 m (0.00) which indicated greater and lower chance of slope failures respectively (Table 2.1).

2.2.2 Slope Character and Landslide Potentiality

The slope map was classified into 10 different slope zones (Fig. 2.1b). Basically, the river basin is attributed with maximum slope angle where slope angle ranges from 0° to more than 89°. The basin is dissected by well developed drainage network which made the slope steepened by continuous branching of the drainage and its headword

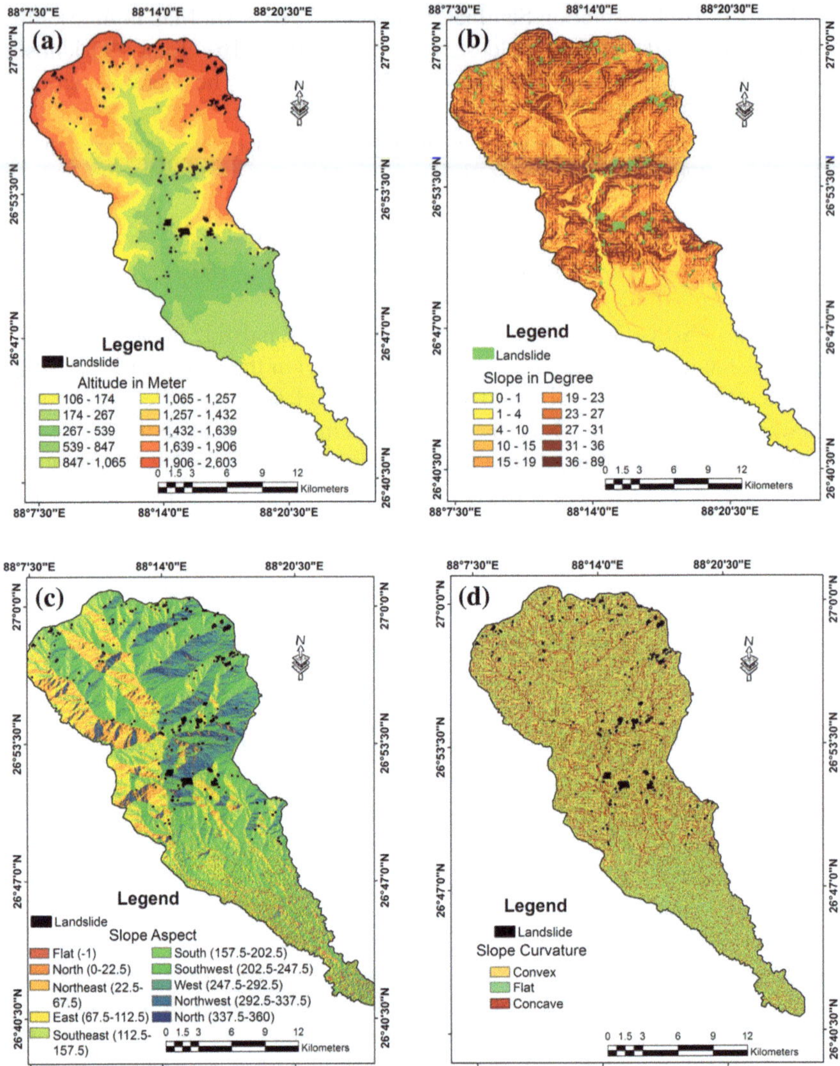

Fig. 2.1 Landslide conditioning factors, **a** altitude, **b** slope, **c** aspect, **d** curvature

erosion. The slope plays an important role for the growth and development of drainage network, surface run-off and soil erosion as well as drainage concentration over the space. Not only that the stream power index (SPI) and topographic wetness index (TWI) are influenced by slope steepness but the study also revealed that there is a positive relationship between slope and landslide frequency ratio. The slope having greater than 27° is attributed with the frequency ratio value ranges from 1.21 to 5.41 which showed high probability of landslide phenomena (Table 2.2).

Table 2.1 Elevation character of the Balason river basin and landslide susceptibility

Factors	Subclasses	Total pixels		Landslide occurrence pixels		FR value
		Absolute	Percentage	Absolute	Percentage	
Altitude (m)	106–174	42,815	10.18	0	0.00	0.00
	174–267	42,258	10.05	0	0.00	0.00
	267–539	41,995	9.99	31	2.27	0.23
	539–847	42,082	10.01	271	19.81	1.98
	847–1065	41,913	9.97	277	20.25	2.03
	1065–1257	42,012	9.99	381	27.85	2.79
	1257–1432	41,959	9.98	87	6.36	0.64
	1432–1639	41,898	9.96	28	2.05	0.21
	1639–1906	41,783	9.94	57	4.17	0.42
	1906–2603	41,785	9.94	236	17.25	1.74

Table 2.2 Slope character of the Balason river basin and landslide susceptibility

Factors	Subclasses	Total pixels		Landslide occurrence pixels		FR value
		Absolute	Percentage	Absolute	Percentage	
Slope (°)	0–1	46,301	11.01	7	0.51	0.05
	1–4	50,128	11.92	13	0.95	0.08
	4–10	40,560	9.65	29	2.12	0.22
	10–15	48,503	11.53	94	6.87	0.60
	15–19	46,043	10.95	102	7.46	0.68
	19–23	46,809	11.13	104	7.60	0.68
	23–27	42,086	10.01	136	9.94	0.99
	27–31	35,846	8.52	141	10.31	1.21
	31–36	31,866	7.58	172	12.57	1.66
	36–89	32,358	7.70	570	41.67	5.41

2.2.3 *Slope Aspect and Landslide Potentiality*

Aspect map was also classified into 10 and prepared using 3D analyst tool (Fig. 2.1c). Slope aspect is an important landslide conditioning parameter because it directionality indicates the exposure to sunlight and drying winds which control the concentration of soil moisture and this soil moisture are positively related to the landslide occurrence (Magliulo et al. 2008). The impact of the slope aspect was evaluated as contributing factor of slope failure. The highest frequency ratio (2.31) was found in the south facing slope and it was lowest (0.02) in the class of north facing slope (Table 2.3). South facing slope of the Balason river basin is exposed to atmospheric

Table 2.3 Slope aspect of the Balason river basin and landslide susceptibility

Factors	Subclasses	Total pixels		Landslide occurrence pixels		FR value
		Absolute	Percentage	Absolute	Percentage	
Slope aspect	Flat (−1)	9005	2.14	4	0.29	0.14
	North (0–22.5)	18,829	4.48	2	0.15	0.03
	Northeast (22.5–67.5)	34,118	8.11	14	1.02	0.13
	East (67.5–112.5)	55,240	13.14	53	3.87	0.29
	Southeast (112.5–157.5)	58,292	13.86	134	9.80	0.71
	South (157.5–202.5)	82,475	19.61	619	45.25	2.31
	Southwest (202.5–247.5)	60,795	14.46	349	25.51	1.76
	West (247.5–292.5)	56,036	13.33	183	13.38	1.00
	Northwest (292.5–337.5)	32,934	7.83	9	0.66	0.08
	North (337.5–360)	12,776	3.04	1	0.07	0.02

processes i.e. rainfall and temperature which plays a significant role in disintegration and decomposition of the surface rocks. South facing slope also receive orographic monsoon rainfall which promote slope saturation and landslides.

2.2.4 *Slope Curvature and Landslide Potentiality*

The prepared curvature map (Fig. 2.1d) was classed into 3 i.e. convex, flat and concave. Negative values represent concave, zero value represents flat and positive values represents convex curvature. Higher the positive and negative value represent the area is more susceptible to slope instability. The study revealed that high positive and negative curvatures are registered with maximum FR value (Table 2.4).

Table 2.4 Slope curvature of the Balason river basin and landslide susceptibility

Factors	Subclasses	Total pixels		Landslide occurrence pixels		FR value
		Absolute	Percentage	Absolute	Percentage	
Slope curvature	Concave	193,016	45.90	707	51.68	1.13
	Flat	34,524	8.21	57	4.17	0.51
	Convex	192,960	45.89	604	44.15	0.96

2.2.5 Geomorphology and Landslide Potentiality

Lama (2003) divided the Balason river basin into 7 distinct geomorphological groups i.e. Upper Hill (>2000 m), Middle Hill (1000–2000 m), Lower Hill (<1000 m), Foothill (Piedmont), Upper Alluvial Fan, Lower Alluvial Fan and River Flood Plain (Fig. 2.2a). Middle Hill (1000–2000 m) covered larger area of the basin (46.22%) while River Flood Plain had the smallest area of the basin (4.07%). The greater potentiality of landslide occurrences was found in lower hills area (<2000 m) (1.90). The chance of slope failures was lowest in the area of Lower Alluvial Fan and River Flood Plain where both were registered with 0.00 FR value (Table 2.5).

2.2.6 Geology and Landslide Potentiality

The geology of Balason river basin was divided into five major categories viz., Darjeeling Gneiss, Daling Series, Damuda Series/Lower Gondwana, Nahan Group and Alluvium (older & recent) (Fig. 2.2b) (Lama 2003). The Darjeeling gneiss covered a large part of the basin (68.33%) and where it varies from a foliated granitoid rock composed to quartz, feldsper and biotite to more or less pure mica schist. The Daling

Table 2.5 Geomorphology of the Balason river basin and landslide susceptibility

Sub-classes of geomorphology	Total pixels		Landslide occurrence pixels		FR value
	Absolute	Percentage	Absolute	Percentage	
Upper hill (>2000 m)	29,309	6.97	150	10.96	1.57
Middle hill (1000–2000 m)	194,339	46.22	732	53.51	1.16
Lower hill (<2000 m)	77,058	18.33	477	34.87	1.90
Foothill (Piedmont)	21,111	5.02	7	0.51	0.10
Upper Alluvial fan	53,984	12.84	2	0.15	0.01
Lower Alluvial fan	27,588	6.56	0	0.00	0.00
River Flood plain	17,111	4.07	0	0.00	0.00

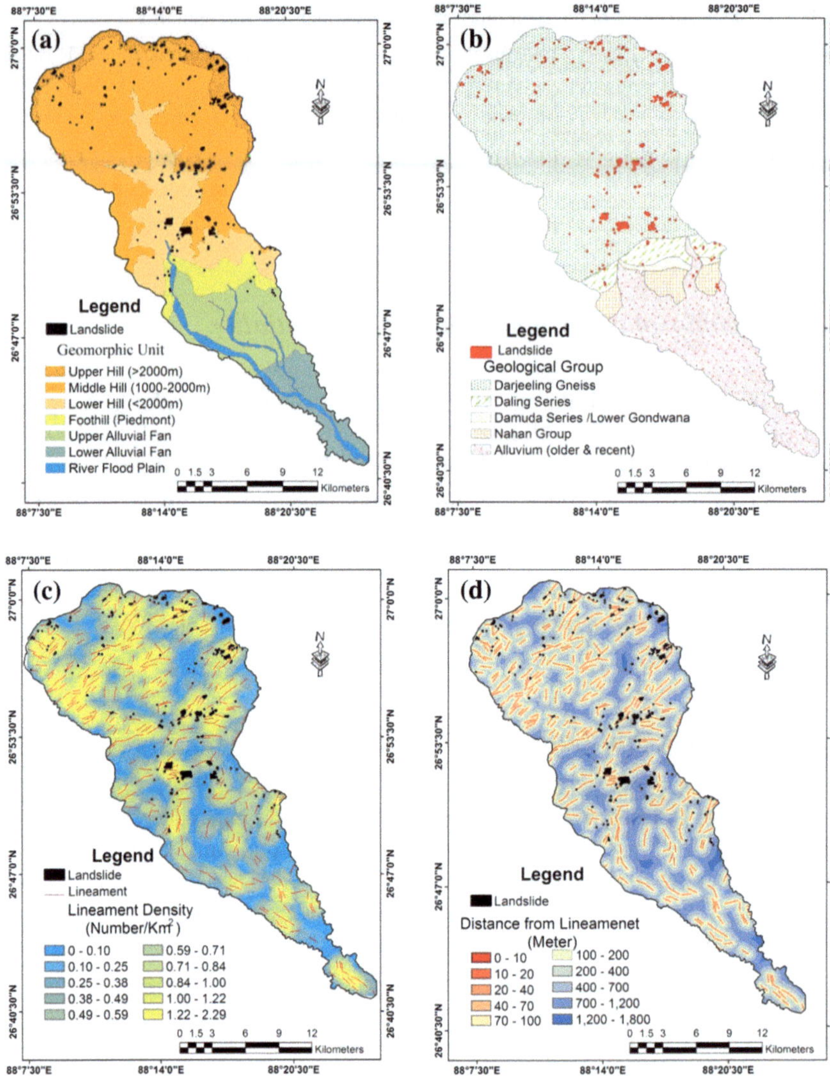

Fig. 2.2 Landslide conditioning factors, **a** geomorphological units, **b** geology, **c** lineament density, **d** distance from lineament

Table 2.6 Geology of the Balason river basin and landslide susceptibility

Factors	Subclasses	Total pixels		Landslide occurrence pixels		FR value
		Absolute	Percentage	Absolute	Percentage	
Geology	Darjeeling Gneiss (Dg)	287,309	68.33	1345	98.32	1.44
	Daling Series (Ds)	18,518	4.40	21	1.54	0.35
	Damuda Series/Lower Gondwana (DAs)	4784	1.14	0	0.00	0.00
	Nahan Group; Lower Siwalik Deposits (Ng)	16,839	4.00	0	0.00	0.00
	Alluvium (older & recent) (Al)	93,050	22.13	2	0.15	0.01

series consists of phyllite, slate and quartzite which are being found in the Middle East and Middle West portion of the basin. Damuda Series/Lower Gondwana series consist of Quartzitic Sandstones, Shales and Slates, Semi-Anthracitic (graphitic) Coal, Lamprophyre Silts and Minor Bands of limestone. The Nahan Group or Lower Siwalik Deposits composed mainly of Soft Greyish Sandstone, Mudstone, Shales and Conglomerates. Alluviums (older & recent) are composed of Boulder Beds and Other Sands and Gravels, Drift Formation and Younger Flood Plain Deposits Comprising Sand & Gravel, Pebbles, etc. It can be seen that, Darjeeling Gneiss had the greater potentiality to landslide occurrences (1.44). Damuda Series/Lower Gondwana and Nahan Group and Lower Siwalik Deposits were registered with lowest FR value (0.00) which indicated low chances of slope failures phenomena (Table 2.6).

2.2.7 *Lineament Density, Distance to Lineament and Landslide Potentiality*

A lineament refers to the expression of underlying linear geological structures such as fault which will comprise a fault-aligned valley, a series of fault or fold-aligned hills, a straight coastline or indeed a combination of these features. The probability of landslide event was highest (2.88) in the areas having highest lineament density. The lowest frequency ratio (0.01) was found at the places where lineament density is low (Fig. 2.2c and Table 2.7). In case of distance to lineament, the highest frequency ratio for distance to lineament classes as main contributors of slope failure belonged to the class of 20–40 m with a ratio of 2.81. The probability of landslide occurrence decreases with the increasing distance from lineament (Fig. 2.2d and Table 2.8).

Table 2.7 Lineament density and landslide susceptibility

Factors	Subclasses	Total pixels		Landslide occurrence pixels		FR value
		Absolute	Percentage	Absolute	Percentage	
Lineament density	0–0.10	39,532	9.40	1	0.07	0.01
	0.10–0.25	43,171	10.27	37	2.68	0.26
	0.25–0.38	43,723	10.40	74	5.42	0.52
	0.38–0.49	43,684	10.39	79	5.77	0.56
	0.49–0.59	41,022	9.76	106	7.76	0.80
	0.59–0.71	43,459	10.34	192	14.07	1.36
	0.71–0.84	42,431	10.09	177	12.90	1.28
	0.84–1.00	45,012	10.70	152	11.12	1.04
	1.00–1.22	39,867	9.48	189	13.80	1.46
	1.22–2.29	38,599	9.18	361	26.42	2.88

Table 2.8 Distance to lineament and landslide susceptibility

Factors	Subclasses	Total pixels		Landslide occurrence pixels		FR value
		Absolute	Percentage	Absolute	Percentage	
Distance to lineament	0–10	5557	1.32	49	3.58	2.71
	10–20	5970	1.42	42	3.07	2.16
	20–40	11,709	2.78	107	7.82	2.81
	40–70	19,252	4.58	154	11.26	2.46
	70–100	19,504	4.64	129	9.43	2.03
	100–200	66,656	15.85	354	25.88	1.63
	200–400	113,338	26.95	239	17.47	0.65
	400–700	110,616	26.31	226	16.52	0.63
	700–1200	62,986	14.98	68	4.97	0.33
	1200–1800	4912	1.17	0	0.00	0.00

2.2.8 *Land Use/Land Cover (LULC) and Landslide Potentiality*

The higher frequency ratio was found in the settlement areas (1.54) and agricultural plantation zone (1.52) which denotes the higher probability of landslide occurrence. The frequency ratio of water bodies and forest area was 0.71 and 0.35 respectively and expected as low probability of landslide occurrence (Fig. 2.3a and Table 2.9).

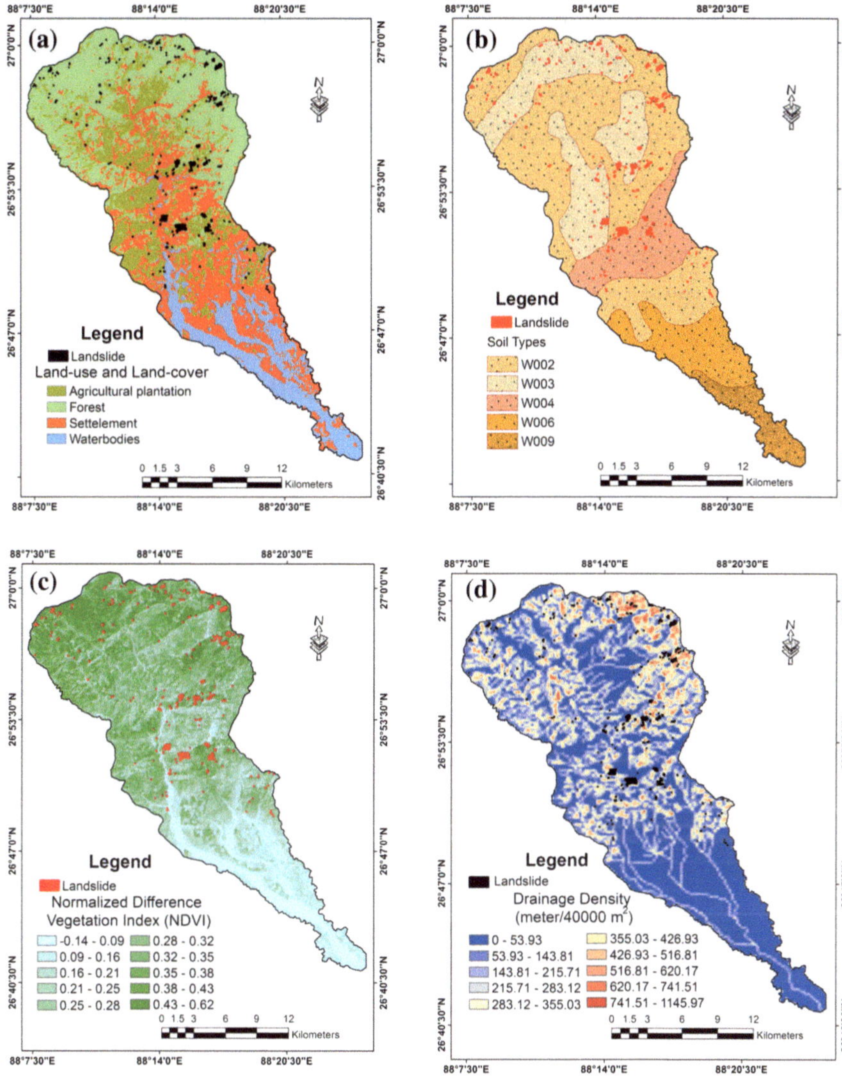

Fig. 2.3 Landslide conditioning factors, **a** land use/land cover, **b** soil type, **c** NDVI and **d** drainage density

Table 2.9 Land use/ land cover (LULC) and landslide susceptibility

Factors	Subclasses	Total pixels		Landslide occurrence pixels		FR value
		Absolute	Percentage	Absolute	Percentage	
Land use/land cover (LULC)	Agricultural plantation	83,710	19.91	413	30.20	1.52
	Water bodies	63,561	15.12	147	10.78	0.71
	Settlement	128,894	30.65	645	47.08	1.54
	Forest	144,335	34.32	163	11.94	0.35

2.2.9 Soil Character and Landslide Probability

Five types of soil were identified in the basin i.e. W002, W003, W004, W006 and W009 (Fig. 2.3b). The soil of the Balason river basin is favorable for tea plantation. The original soil of the area is drastically changed either by the process of new soil formation on the truncated top or by removal of soil by erosion (Sarkar 1990). The parent soil of this region has changed due to its nature and it has formed more than one time (Lama 2003). In the sandy area, soils are generally siliceous and aluminous. The lower Balason basin usually formed with fine grained clay loam. In some cases the top soils are finer in nature than the subsurface soils. The detailed characteristics of each type of soil of the Balason river basin were presented in Table 2.10. In this basin, W003 soil type was registered with highest frequency ratio is 1.88. On the other hand, W006 and W009 soil types covering almost 17% area of the basin have the lowest chance of landslide events (Table 2.10).

2.2.10 NDVI and Landslide Probability

The positive values indicate vegetation health and negative values represent water depth. In this study, ten classes have been made using quantile classification method (Fig. 2.3c). The highest frequency ratio (2.10) was found in the class of 0.16–0.21 which depicts that the slopes exposed to atmospheric process and make the slope more susceptible to slope instability. The frequency ratio value was lowest (0.26) for the class of 0.43–0.62 (Table 2.11).

2.2.11 Drainage Density and Landslide Susceptibility

The drainage network development on both sides of the slope segments make the interfluves area more steep resulting mountain slope instability. The drainage network development, slope steepening, and slope failure are regular processes which

Table 2.10 Soil characteristics of Balason river basin

Symbol	Soil characteristics	Taxonomic name	Frequency ratio
W002	Moderately shallow, excessively drained, coarse loamy soils occurring on steep side slopes with gravelly loamy surface, severe erosion and strong rockiness Associated with moderately shallow, well drained, gravelly loamy soils with loamy surface and moderate erosion	Coarse loamy, typic Udorthents Loamy skeletal, typic Dystrochrepts	0.81
W003	Deep, well drained, fine loamy soils occurring on steep side slopes with gravelly loamy surface, moderate erosion and moderate rockiness Associated with moderately shallow, excessively drained, coarse loamy soils with loamy surface, severe erosion and moderate rockiness	Fine loamy, Umbric Dystrochrepts Coarse loamy, Typic Udorthents	1.88
W004	Moderately shallow, well drained, gravelly loamy soils occurring on steep side slopes with gravelly loamy surface, moderate erosion and moderate rockiness Associated with moderately shallow, somewhat excessively drained, gravelly loamy soil with loamy surface, moderate erosion and moderate rockiness	Loamy skeletal, Typic Haplumbrepts. Loamy skeletal, typic Udorthents	1.36
W006	Very deep, imperfectly drained, coarse loamy soils occurring on very gently sloping upper piedmont plains with loamy surface and moderate erosion Associated with very deep, imperfectly drained fine loamy soils	Coarse loamy, Umbric Dystrochrepts Fine loamy, Fluventic Dystrochrepts	0.00
W009	Very deep, imperfectly drained, coarse loamy soils occurring on nearly level lower piedmont plain with loamy surface associated with very deep imperfectly drained, fine loamy soils	Coarse loamy, Aquic Udifluvents Fine loamy, Fluventic Eutrochrepts	0.00

Table 2.11 NDVI and landslide susceptibility

Factors	Subclasses	Total pixels		Landslide occurrence pixels		FR value
		Absolute	Percentage	Absolute	Percentage	
NDVI	−0.14 to 0.09	40,463	9.62	94	6.86	0.71
	0.09–0.16	44,788	10.65	148	10.84	1.02
	0.16–0.21	41,014	9.75	279	20.45	2.10
	0.21–0.25	44,540	10.59	277	20.25	1.91
	0.25–0.28	48,426	11.52	222	16.20	1.41
	0.28–0.32	41,256	9.81	116	8.51	0.87
	0.32–0.35	41,803	9.94	79	5.77	0.58
	0.35–0.38	35,835	8.52	54	3.91	0.46
	0.38–0.43	45,093	10.72	67	4.87	0.45
	0.43–0.62	37,282	8.87	32	2.33	0.26

Table 2.12 Drainage density of the Balason river basin and landslide susceptibility

Factors	Subclasses	Total pixels		Landslide occurrence pixels		FR value
		Absolute	Percentage	Absolute	Percentage	
Drainage density	0–53.93	119,003	28.30	218	15.94	0.56
	53.93–143.81	97,468	23.18	380	27.78	1.20
	143.81–215.71	76,554	18.21	274	20.03	1.10
	215.71–283.12	55,176	13.12	218	15.94	1.21
	283.12–355.03	37,793	8.99	131	9.58	1.07
	355.03–426.93	18,793	4.47	39	2.85	0.64
	426.93–516.81	9641	2.29	39	2.85	1.24
	516.81–620.17	4674	1.11	48	3.51	3.16
	620.17–741.51	1178	0.28	19	1.39	4.96
	741.51–1145.97	220	0.05	2	0.15	2.92

promotes equilibrium in the geomorphic system. The concentration of drainage network and their engagement in the process of erosion and transportation make the slope more vulnerable to landslip phenomena. The drainage density varies from one place to another based on the developed drainage network in the basin. Higher the drainage density more is the drainage concentration and slope saturation over the space. The drainage density map was divided into 10 classes (Fig. 2.3d). It was observed that, high drainage density of more than 2 covered about 1.5% area of the basin. 620.17–741.51 class having maximum chance of landslide occurrences (4.96) whereas 0–53.93 drainage density class having least chance of slope failures (0.00) (Table 2.12).

Table 2.13 Distance to drainage and landslide susceptibility

Factors	Subclasses	Total pixels		Landslide occurrence pixels		FR value
		Absolute	Percentage	Absolute	Percentage	
Distance to drainage	0–10	41,936	9.97	172	12.60	1.26
	10–20	20,857	4.96	91	6.67	1.34
	20–40	55,757	13.26	214	15.64	1.18
	40–70	65,061	15.47	227	16.60	1.07
	70–100	47,405	11.27	188	13.71	1.22
	100–150	53,424	12.70	247	18.01	1.42
	150–300	69,150	16.44	227	16.60	1.01
	300–700	43,921	10.45	2	0.15	0.01
	700–1500	19,226	4.57	0	0.00	0.00
	1500–2500	3763	0.89	0	0.00	0.00

2.2.12 Distance to Drainage and Landslide Potentiality

To estimate distance to drainage, 10 buffer classes (10, 20, 40, 70, 100, 150, 300, 700, 1500, 2500 m) were made using multiple buffer tool in Arc Map 10.1 (Fig. 2.4a). The high frequency ratio value was found in the class of 100–150 m (1.42), 10–20 m (1.34), 0–10 m (1.26), 70–100 m (1.22), 20–40 m (1.18), 40–70 m (1.07) and 150–300 m (1.01) indicating greater chance of slope failure respectively. Rest of the classes have very low to zero frequency ratios ranging from 0.01–0.00 (Table 2.13).

2.2.13 Stream Power Index (SPI), Topographic Wetted Index (TWI) and Landslide Potentiality

SPI is the measure of erosive power of the overland flow (Moore et al. 1991) and TWI refers to the accumulation of flow at any point within a watershed. The frequency ratio was <1 except when the SPI was ranges from the class of 11.67–12.34 to 14.47–20.21 which indicates high probability of slope instability. The highest (1.95) and lowest (0.08) ratio was found in the class of 12.34–13.14 and −13.82 to 1.93 respectively (Fig. 2.4b and Table 2.14). The ratio was highest (2.16) when TWI was −19.41 to −17.59 indicating high chance to landslide occurrence. The chance of occurrence gradually decreases from the class of −9.92 to −9.08 to −5.87 to 16.19. The lowest frequency (0.07) found in the class of −16.62 to −14.66 (Fig. 2.4c and Table 2.15).

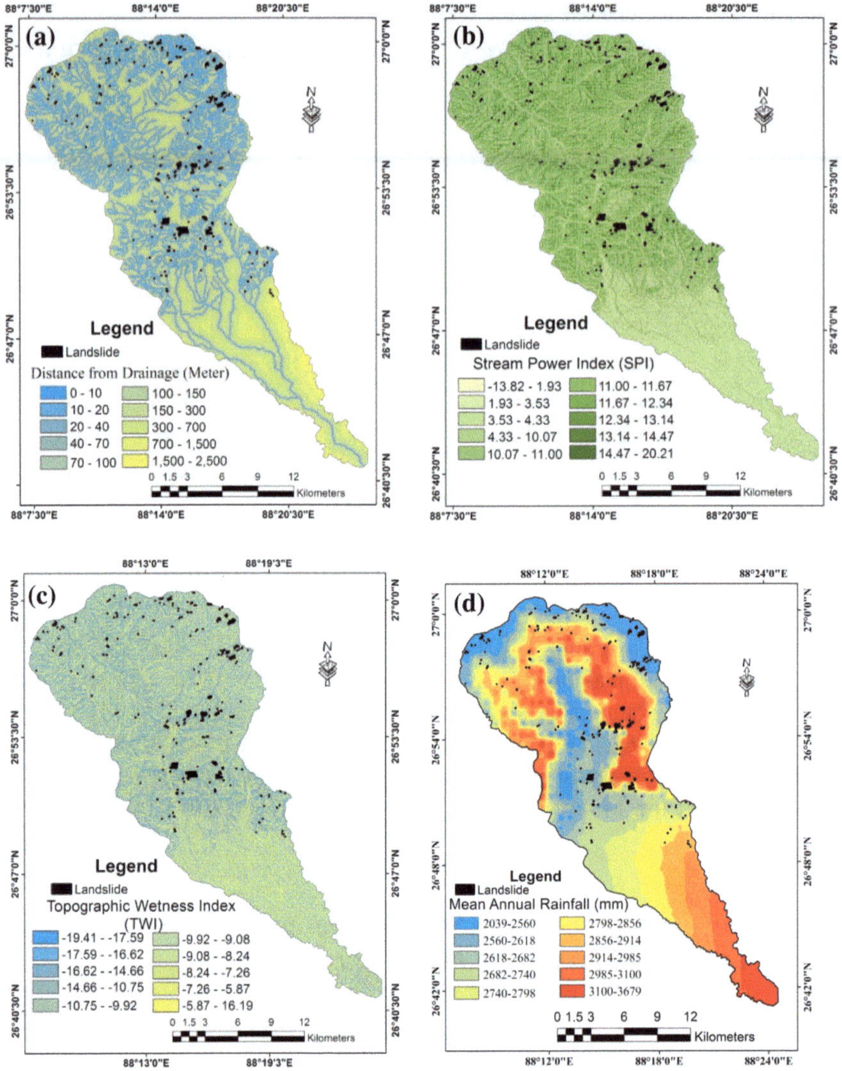

Fig. 2.4 Landslide conditioning factors, **a** distance to drainage, **b** stream power index, SPI, **c** topographic wetness index, TWI and **d** rainfall

Table 2.14 Stream power index (SPI) and landslide susceptibility

Factors	Subclasses	Total pixels		Landslide occurrence pixels		FR value
		Absolute	Percentage	Absolute	Percentage	
Stream power index (SPI)	−13.82 to 1.93	40,628	9.66	10	0.73	0.08
	1.93–3.53	47,727	11.35	37	2.70	0.24
	3.53–4.33	46,419	11.04	157	11.48	1.04
	4.33–10.07	44,636	10.61	127	9.28	0.87
	10.07–11.00	42,730	10.16	45	3.29	0.32
	11.00–11.67	47,117	11.20	141	10.31	0.92
	11.67–12.34	41,514	9.87	176	12.87	1.30
	12.34–13.14	38,179	9.08	242	17.69	1.95
	13.14–14.47	37,741	8.98	238	17.40	1.94
	14.47–20.21	33,809	8.04	195	14.25	1.77

Table 2.15 Topographic wetted index (TWI) and landslide susceptibility

Factors	Subclasses	Total pixels		Landslide occurrence pixels		FR value
		Absolute	Percentage	Absolute	Percentage	
Topographic wetted index (TWI)	−19.41 to −17.59	35,280	8.39	248	18.12	2.16
	−17.59 to −16.62	49,004	11.65	63	4.59	0.39
	−16.62 to −14.66	42,254	10.05	10	0.73	0.07
	−14.66 to −10.75	42,590	10.13	241	17.61	1.74
	−10.75 to −9.92	45,326	10.78	184	13.46	1.25
	−9.92 to −9.08	44,372	10.55	188	13.76	1.30
	−9.08 to −8.24	40,495	9.63	135	9.90	1.03
	−8.24 to −7.26	40,150	9.55	127	9.32	0.98
	−7.26 to −5.87	41,303	9.82	111	8.08	0.82
	−5.87 to 16.19	39,726	9.45	61	4.44	0.47

Table 2.16 Rainfall of the Balason river basin and landslide susceptibility

Factors	Subclasses	Total pixels		Landslide occurrence pixels		FR value
		Absolute	Percentage	Absolute	Percentage	
Rainfall (mm)	2039–2560	40,635	9.66	130	9.48	0.98
	2560–2618	42,137	10.02	209	15.25	1.52
	2618–2682	44,545	10.59	186	13.58	1.28
	2682–2740	38,867	9.24	161	11.80	1.28
	2740–2798	45,248	10.76	186	13.58	1.26
	2798–2856	41,930	9.97	147	10.72	1.08
	2856–2914	41,695	9.92	87	6.36	0.64
	2914–2985	42,379	10.08	40	2.91	0.29
	2985–3100	42,956	10.22	62	4.53	0.44
	3100–3679	40,108	9.54	161	11.80	1.24

2.2.14 *Spatial Distribution of Rainfall and Landslide Potentiality*

Landslides are closely associated with duration, amount and intensity of rainfall. The intensity and amount of rainfall varies from one place to another in the basin due to topographic configuration. The mean annual rainfall ranges from 2041 to 3673 mm. Based on the rainfall distribution, the Balason river basin was classified into ten zones (Fig. 2.4d). Around 35% landslide affected area and 50% area of basin are characterized by mean annual rainfall of more than 2790 mm. More than 1.00 frequency ratio value was found in the class ranges from 2559 to 2847 mm and 3090–3673 rainfall which was characterized by high landslide events. Rest of the class having less than 1.00 frequency ratio value indicating minimum chance of landslide occurrences. Some high annual rainfall classes i.e. 2847–2905 mm, 2905–2975 mm and 2975–3090 mm have low FR value because of the unfavorable geo-environmental conditions in the concerned area (Table 2.16).

2.3 Conclusion

All the geomorphic parameters play a significant role in causing landslides in the Balason river basin. Not only the factors but also the individual category of all the landslide occurrence factors showed a distinct level of landslide probability. The experiences about the concerned study area revealed that the amount and intensity of rainfall, drainage network, lithological composition, soil character and slope character plays a significant role to cause landslide events in the Darjeeling Himalaya. It is finally assumed that the results will help the policy makers, developers and engineers for choosing suitable locations to implement slope management plans, land use plans and other development action plans.

References

Devkota KC, Regmi AD, Pourghasemi HR et al (2013) Landslide susceptibility mapping using certainty factor, index of entropy and logistic regression models in GIS and their comparison at Mugling-Narayanghat road section in Nepal Himalaya. Nat Hazards 65(1):135–165. https://doi.org/10.1007/s11069-012-0347-6

Grozavu A, Plescan S, Patriche CV, Margarint MC, Rosca B (2013) Landslide susceptibility assessment: GIS application to a complex mountainous environment, the carpathians: integrating nature and society towards sustainability, environmental science and engineering, pp 31–44

Karim S, Jalileddin S, Ali MT (2011) Zoning landslide by use of frequency ratio method (case study: Deylaman Region) Middle-East. J Sci Res 9(5):578–583

Lama IL (2003) Study of the environmental geomorphology in the Balason Basin, unpublished Ph. D thesis, North Bengal University, pp 24–28

Lepcha I (2012) Regional study of geo-environmental aspects: a study of Balason Basin, Darjeeling district, West Bengal. J Geo-Environ Observer 1(2):24–37

Mandal S, Maiti R (2014) Semi-quantitative approaches for landslide assessment and prediction, Part of the series Springer natural hazards

Magliulo P, Di Lisio A, Russo F, Zelano A (2008) Geomorphology and landslide susceptibility assessment using GIS and bivariate statistics: a case study in Southern Italy. Nat Hazards 47:411–435

Moore ID, Grayson RB, Ladson AR (1991) Digital terrain modeling—a review of hydro-hydrological geomorphological and biological application. Hydrol Process 5:3–30

Nautiiyal SP (1966) On the stability of certain hill slopes in and around Darjeeling WB. Bull Geol Surv India 15:31–48

Sarkar S (1990) Genesis and classification of the soils of the Mahananda basin, Darjeeling Himalaya. Geog Mem 2(182):117–129

Sengupta CK (2000) Report on landslide hazard microzonation of Mirik Municipal area for Urban Planning. Unpublished GSI report FS 1999–2000

Tamang L (2013) Effects of boulder lifting on the fluvial characteristics of lower Balason basin in Darjeeling district west Bengal, Ph.D. thesis, University of North Bengal, Department of Geography and Applied Geography, p 19

Umar Z, Pradhan B, Ahmad A et al (2014) Earthquake induced landslide susceptibility mapping using an integrated ensemble frequency ratio and logistic regression models in West Sumatera Province, Indonesia. Catena 118:124–135. https://doi.org/10.1016/j.catena.2014.02.005

Chapter 3
Slope Instability Analysis Using Morphometric Parameters: A Sub-watersheds Scale Study

3.1 Introduction

Drainage basin or watershed is the basic or ideal areal unit of the earth surface in geomorphological study of fluvially eroded landscapes (Chorley 1969), simply for the management and sustainable development of natural resources (Patel et al. 2013). Keeping in view of the increasing population demand and food security for the future generation, it has been realized that the water and land resources need to be developed, used and managed comprehensively in an integrated way (Biswas et al. 1999). Morphometric analysis is very much important in analyzing the changing quality of water and soil of a small hilly micro-watershed. In addition, it is very useful in regional flood frequency analysis, hydrological modeling, watershed prioritization, natural resources conservation and management, drainage basin evaluation, etc. For that, it requires measurement of linear, areal and relief aspects of the basin (Nautiyal 1994). The estimation of morphometric parameters of a hilly catchment gains more importance in spite of the difficulty in collection and measurement of hydrologic data. Hence, GIS is an effective tool not only for collection, storage, management and retrieval of spatial and non-spatial data, but also for the derivation of the useful results and modeling (Gupta and Srivastava 2010; Srivastava et al. 2011). Pioneering work on the drainage basin morphometry has been carried out by Horton (1932, 1945), Smith (1950), Strahler (1952, 1964), Miller (1953), Schumm (1956), Melton (1958) and Chorley (1969). Prioritization of sub-watersheds based on morphometric analysis were attempted by Biswas et al. (1999), Chopra et al. (2005), Patel et al. (2013), Rekha et al. (2011), Panhalkar and Pawar (2011), Vandana (2012), Patil and Mali (2013), Panda and Nagarajan (2013), Tolessa and Rao (2013), Ali and Ali (2014), Amani and Safaviyan (2015) etc. Shrimali et al. (2011) delineated and prioritized of soil erosion areas in Sukhana lake catchment in the Shiwalik hills by using RS and GIS. Nookaratnam et al. (2005) carried out check dam positioning by prioritization of micro-watersheds using Sediment Yield Index (SYI) model and

S. Mandal and S. Mondal, *Geoinformatics and Modelling of Landslide Susceptibility and Risk*, Environmental Science and Engineering,
https://doi.org/10.1007/978-3-030-10495-5_3

morphometric analysis using GIS. Javed et al. (2009, 2011) studied on prioritization of sub-watersheds of Kanera watershed in Guna district of Madhya Pradesh and Jaggar watershed in the Karauli district, Rajasthan respectively based on morphometric and land use analysis using RS and GIS techniques. Patel et al. (2013) have carried out a case study to select suitable sites for water harvesting structures in Varekhadi watershed, a part of Lower Tapi Basin (LTB), Surat district, Gujarat by overlaying of Shuttle Radar Topography Mission DEM, soil map and slope map using RS and GIS approach. Chirala et al. (2015) worked on Meghadrigedda catchment in Visakhapatnam, India for mapping of soil erosion zones using RS and GIS techniques. The present study focuses on sub-watersheds wise slope instability analysis and prioritization based on the 13 morphometric parameters of the Balason river basin, Darjeeling Himalaya, West Bengal, India, using RS and GIS techniques. Prioritization rating of 18 sub-watersheds is carried out through compound ranking method. The sub-watershed with the lowest rank is given the highest priority in terms of erosion and suggested for conservation measures urgently.

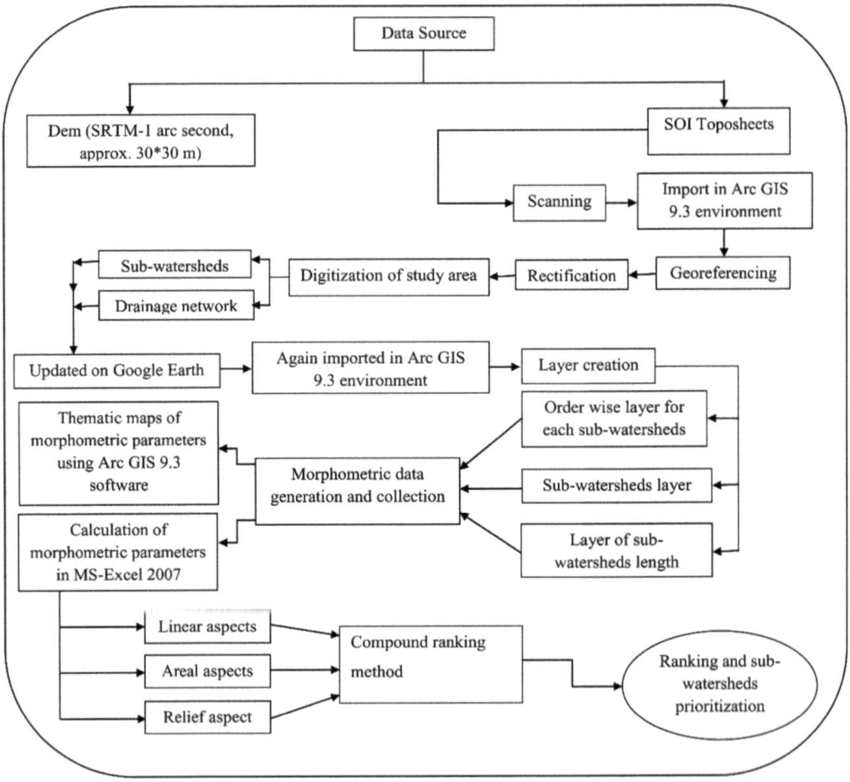

Fig. 3.1 Methodology of the study

3.2 Materials and Methods

The present study was basically based on morphometric analysis of the Balason river basin of Darjeeling Himalaya. To assess the morphometric characters, SRTM DEM (30 m resolution), Survey of India (SOI) Topo-sheets and Geological map were used. The drainage network was initially derived from SOI toposheets 78A/4, 78A/8, 78B/1, 78B/5 and 78B/6 (Scale 1:50,000) and later updated using Google Earth. The sub-watersheds boundaries were demarcated on the basis of slope, elevation, 3D view and drainage flow directions and the basin was divided into 18 sub-watersheds. To find out the sub-watersheds which are more unstable in regard to slope instability two major steps have been considered i.e. (i) *assessment of morphometric parameters* and *preparation of spatial data layers* and (ii) *slope instability analysis* and *sub-watersheds wise prioritization using compound ranking method* (Fig. 3.1).

3.2.1 Assessment of Morphometric Parameters and Preparation of Spatial Data Layers

The morphometric parameters such as bifurcation ratio (Rb), length of overland flow (Lof), drainage density (Dd), stream frequency (Fs), texture ratio (Rt), drainage texture (Td), compactness coefficient (Cc), constant of channel maintenance (Ccm), shape factor/basin shape (Sf), form factor (Ff), circularity ratio (Rc), elongation ratio (Re) and relief ratio (Rr) were computed using standard methods and formulae and prepared data layers in GIS (Table 3.1).

3.2.1.1 Bifurcation Ratio (Rb)

Rb is an index of reliefs and dissections (Horton 1945). It is a dimensionless property that expresses the ratio of the number of streams of a given order to the number of streams of the next higher order (Horton 1932, 1945; Schumm 1956). In different environmental condition as well as different regions Rb values shows only a little variation except where very strong geological control dominates (Strahler 1957) and morphometric control of the basin itself (Singh et al. 1984). Lower Rb values are the results of structurally less disturbed watershed without any distortion in drainage pattern (Javed et al. 2009; Magesh et al. 2012) and are the attributed to the relatively gentle slope (Panda and Nagarajan 2013). In this study, Weighted Mean Bifurcation Ratio (Rbwm) was considered which obtained by multiplying the bifurcation ratio for each successive pair of orders by the total numbers of streams involved in the ratio and taking the mean of the sum of these values.

Table 3.1 Empirical formula for calculating of morphometric parameters

Sl. No.	Morphometric parameters	Types/ aspects	Methods/ formulae	References	Units	Results
1	Bifurcation ratio (Rb)	Linear aspect	Rb = Nu/Nu + 1 Nu = Total number of stream segment of order 'u', Nu + 1 = Number of stream segment of next higher order	Schumm (1956)	Unit less	1.6–4.103
2	Length of overland flow (Lof)	Linear aspect	Lof = 1/2Dd Dd = Drainage density	Horton (1945)	km/km^2	0.069–0.857
3	Drainage density (Dd)	Areal aspect	Dd = L/A L = Total length of streams, A = Area of watershed	Horton (1945)	km/km^2	0.583–7.207
4	Stream frequency (Fs)	Areal aspect	Fs = N/A N = Total number of streams, A = Area of watershed	Horton (1945)	No. of $stream/km^2$	0.069–29.447
5	Texture ratio (Rt)	Areal aspect	Rt = N1/P N1 = Total number of first order streams, P = Perimeter of watershed	Horton (1945), Schumm (1956)	No. of first order stream/km	0–16.958
6	Drainage texture (Td)	Areal aspect	Td = Nu/P Nu = Total number of streams, P = Perimeter of watershed	Horton (1945)	No. of stream/km	0.069–23.158

(continued)

Table 3.1 (continued)

Sl. No.	Morphometric parameters	Types/ aspects	Methods/ formulae	References	Units	Results
7	Compactness coefficient (Cc)	Areal aspect	Cc = 0.2841*P/$A^{0.5}$ P = Perimeter of watershed, A = Area of watershed	Gravelius (1914)	Unit less	1.198–2.156
8	Constant of channel maintenance (Ccm)	Areal aspect	Ccm = 1/Dd Dd = Drainage density	Horton (1945)	km^2/km	0.139–18.714
9	Shape factor/basin shape (Sf)	Areal aspect	Sf = Lb^2/A Lb = Length of watershed, A = Area of watershed	Horton (1932)	Unit less	1.780–5.804
10	Form factor (Ff)	Areal aspect	Ff = A/Lb^2 A = Area of watershed, Lb = Length of watershed	Horton (1932)	Unit less	0.172–0.562
11	Circulatory ratio (Rc)	Areal aspect	Rc = 4ЛA/P^2 A = Area of watershed, Л = 3.14, P = Perimeter of watershed	Miller (1953)	Unit less	0.218–0.707
12	Elongation ratio (Re)	Areal aspect	2√(A/Л)/Lb A = Area of watershed, Л = 3.14, Lb = Length of watershed	Schumm (1956)	Unit less	0.468–0.846
13	Relief Ratio (Rr)	Relief aspect	Rr = Bh/Lb Bh = Basin relief, Lb = Basin length	Schumm (1956)	Enumerative	0.023–0.302

3.2.1.2 Length of Overland Flow (Lof)

Lof is one of the most independent variables affecting both the hydrologic and physiographic development of drainage basin (Sharma and Padmaja 1982). It refers to the length of the run of the rainwater on the ground surface before it localized into definite channels and he considered length of overland flow is about half the distance between stream channels and had taken it to be roughly equal to the half the reciprocal of the drainage density (Horton 1945). Length of overland flow is inversely relates with the average channel slope (Patel et al. 2013) (Table 3.2).

3.2.1.3 Drainage Density (Dd)

It indicates the closeness of spacing between channels (Javed et al. 2009) and is a measure of the stream length per unit area in a region of watershed (Horton 1932, 1945; Strahler 1952; Melton 1958); thus ‘Dd’ has units of the reciprocal of length (Patel et al. 2013). Structures, relief history (Pareta and Pareta 2012), resistance to weathering, permeability of rock formation, climate, and vegetation etc. are the important controlling factors of drainage density (Javed et al. 2009). ‘Dd’ is helpful to study hydrologic response to rainfall events (Melton 1957; Gupta and Srivastava 2010), landscape dissection, runoff potential, infiltration capacity, climatic condition and vegetation cover of the basin (Patil and Mali 2013; Ali and Ali 2014). Singh (2010) divided drainage density into five categories i.e. (i) very low (ii) low (iii) moderate (iv) high (v) very high.

3.2.1.4 Stream Frequency (Fs)

The word ‘stream frequency’ introduced by Horton refers to the total number of stream segments of all orders per unit area (Horton 1932, 1945). It has the positive correlation with ‘Dd’ indicating the increase in stream frequency with respect to increase density (Javed et al. 2009; Patel et al. 2013; Vandana 2012; Parveen et al. 2012; Tolessa and Rao 2013). ‘Fs’ can be classified into five categories i.e. very poor, poor, moderate, high and very high (Singh 2010).

3.2.1.5 Texture Ratio (Rt)

Texture ratio is an important factor in the drainage morphometric analysis which depends on the underlying lithology, infiltration capacity and relief aspects of the terrain and it can be defined as the ratio between first order stream streams and perimeter of the basin (Schumm 1956).

Table 3.2 Results of some morphometric parameters

Watershed no.	Basin area (A) in km^2	Basin Perimeter (P) in km	No. of first order stream	Total stream length	Total number of stream	Drainage density in km/km^2	Drainage frequency in no. of stream/km^2	Texture ratio in no. of first order stream/km	Drainage texture in no. of stream/km	Constant channel mainte-nance	Length of overland flow
SW 1	54.21	36.40	442	264.95	592	4.887	10.920	12.143	16.264	0.205	0.102
SW 2	13.28	16.55	82	52.88	110	3.982	8.283	4.955	6.647	0.251	0.126
SW 3	10.47	14.19	117	52.27	155	4.992	14.804	8.245	10.923	0.200	0.100
SW 4	5.69	10.62	94	37.20	131	6.538	23.023	8.851	12.335	0.153	0.076
SW 5	14.84	18.87	320	106.95	437	7.207	29.447	16.958	23.158	0.139	0.069
SW 6	28.96	24.17	264	143.06	373	4.940	12.880	10.923	15.432	0.202	0.101
SW 7	24.44	37.52	138	93.23	186	3.815	7.610	3.678	4.957	0.262	0.131
SW 8	12.79	19.56	138	76.56	191	5.986	14.934	7.055	9.765	0.167	0.084
SW 9	29.86	27.00	431	188.60	597	6.316	19.993	15.963	22.111	0.158	0.079
SW 10	11.25	16.77	60	39.31	84	3.494	7.467	3.578	5.009	0.286	0.143
SW 11	20.50	25.63	109	74.34	152	3.626	7.415	4.253	5.931	0.276	0.138
SW 12	7.69	11.69	43	28.04	59	3.646	7.672	3.678	5.047	0.274	0.137
SW 13	18.67	29.18	129	64.93	183	3.478	9.802	4.421	6.271	0.288	0.144
SW 14	19.31	18.60	32	37.04	44	1.918	2.279	1.720	2.366	0.521	0.261
SW 15	27.57	29.96	185	100.93	258	3.661	9.358	6.175	8.611	0.273	0.137
SW 16	14.00	19.70	101	53.93	151	3.852	10.786	5.127	7.665	0.260	0.130
SW 17	6.74	12.35	59	28.59	83	4.242	12.315	4.777	6.721	0.236	0.118
SW 18	58.18	57.65	0	33.94	4	0.583	0.069	0.000	0.069	1.714	0.857

3.2.1.6 Drainage Texture (Td)

Drainage texture is an important geomorphic concept which means that the relative spacing of drainage lines (Smith 1950). Horton (1945) defined drainage texture as the total number of stream segments of all orders per perimeter of the area. The drainage density and drainage frequency have been collectively defined as drainage texture (Tolessa and Rao 2013). According to Singh (2010) drainage texture refers to the relative spacing of streams per unit length in grid squares. It depends on the underlying lithology, infiltration capacity and relief aspects of the terrain. Smith (1950) and Singh (2010) classified drainage texture into five different textures i.e. very coarse (<2), coarse (2–4), moderate (4–6), fine (6–8) and very fine (>8).

3.2.1.7 Compactness Coefficient (Cc)

Cc can be defined as the ratio of perimeter of watershed to circumference of circular area, which equals the area of the watershed (Gravelius 1914) and also known as Gravelius Index (GI). A circular basin will yield the shortest time of concentration before peak flow occurs in the basin, thus it is most hazardous from a drainage point of view (Javed et al. 2009; Nookaratnam et al. 2005). It is dependent only slope. This parameter is inversely related with the elongation ratio (Table 3.3).

3.2.1.8 Constant of Channel Maintenance (Ccm)

Constant of Channel Maintenance is an important property of landforms and it can be defined as the inverse of drainage density (Schumm 1956). It means that the number of square km or kms of basin surface required developing and sustaining of a channel of 1 km long. According to Strahler (1957) it indicates the relative size of landforms units in a drainage basin and has a specific genetic connotation. The values of Ccm are divided into five classes i.e. very high, high, moderate, low and very low.

3.2.2 Shape Factor/Basin Shape (Sf)

Sf is the ratio of the square of the basin length to area of the basin (Horton 1945) and inversely related with form factor. Lower value of Sf reveals peaked flood discharge while higher value indicates weaker flood discharge periods.

3.2.2.1 Form Factor (Ff)

According to Horton (1932, 1945) form factor may be defined as the ratio of basin area to square of the basin length. For a perfectly circular watershed the value of

Table 3.3 Compactness co-efficient, basin shape, relative relief and relief ratio

Compactness coefficient (Cc)	Basin shape (Sb)	Longest axis in km	Highest elevation (m)	Lowest elevation (m)	Relative relief (rr)	Relief ratio (Rr)
1.40	3.28	13.33	2300	580	1720	0.129
1.29	2.83	6.13	2300	1020	1280	0.209
1.25	2.31	4.92	2300	1080	1220	0.248
1.26	2.97	4.11	2320	1080	1240	0.302
1.39	3.47	7.18	2460	940	1520	0.212
1.28	2.94	9.22	2613	820	1793	0.194
2.16	3.43	9.16	1720	540	1180	0.129
1.55	5.56	8.43	2420	660	1760	0.209
1.40	2.84	9.21	2420	540	1880	0.204
1.42	2.95	5.76	2000	520	1480	0.257
1.61	1.78	6.04	1680	400	1280	0.212
1.20	2.55	4.43	1400	420	980	0.221
1.92	5.80	10.41	1920	200	1720	0.165
1.20	3.00	7.61	980	200	780	0.102
1.62	4.01	10.52	1240	200	1040	0.099
1.50	4.94	8.32	1820	220	1600	0.192
1.35	3.97	5.17	1120	220	900	0.174
2.15	5.03	17.10	500	105	395	0.023

form factor would always be less than 0.754 (Chopra et al. 2005; Javed et al. 2009; Pareta and Pareta 2012). Smaller the value of form factor, more elongated will be the basin.

3.2.2.2 Circularity Ratio (Rc)

Quantitative dimensionless circularity ratio is used for the outline form of a watershed (Miller 1953; Strahler 1964). It can be defined as the ratio of the area of a basin to the area of a circle having the same circumference as the perimeter of the basin (Miller 1953). It depends on the length and frequency of streams, geological structures, land use/land cover, climate and slope of the basin. Rc varies from 0 (a line) to 1 (a circle).

3.2.2.3 Elongation Ratio (Re)

Elongation ratio is defined as the ratio of diameter of a circle of the same area as the basin to the maximum basin length (Schumm 1956). This ratio runs between 0.6 and

Table 3.4 Form factor, circulatory ratio and elongation ratio

Watershed no.	Area (A) in km^2	Length in km (Lb)	Perimeter (P) in km	Form factor (A/Lb^2)	Circulatory ratio (4ΠA/P^2)	Elongation ratio 2√(A/Π)/Lb
1	54.21	13.33	36.40	0.305	0.514	0.623
2	13.28	6.13	16.55	0.353	0.609	0.671
3	10.47	4.92	14.19	0.433	0.653	0.742
4	5.69	4.11	10.62	0.337	0.634	0.655
5	14.84	7.18	18.87	0.288	0.523	0.605
6	28.96	9.22	24.17	0.341	0.623	0.659
7	24.44	9.16	37.52	0.291	0.218	0.609
8	12.79	8.43	19.56	0.180	0.420	0.479
9	29.86	9.21	27.00	0.352	0.514	0.669
10	11.25	5.76	16.77	0.339	0.502	0.657
11	20.50	6.04	25.63	0.562	0.392	0.846
12	7.69	4.43	11.69	0.392	0.707	0.706
13	18.67	10.41	29.18	0.172	0.275	0.468
14	19.31	7.61	18.60	0.333	0.701	0.652
15	27.57	10.52	29.96	0.249	0.386	0.563
16	14.00	8.32	19.70	0.202	0.453	0.507
17	6.74	5.17	12.35	0.252	0.555	0.567
18	58.18	17.10	57.65	0.199	0.220	0.503

1.0 over a wide variety of climatic and geologic types and it can be classified into five categories i.e. circular (0.9–1.0), oval (0.8–0.9), less elongated (0.7–0.8), elongated (0.5–0.7) and more elongated (<0.5) (Strahler 1964). The value of Re varies from 0 (highly elongated shape) to 1 (circular shape).

3.2.2.4 Relief Ratio (Rr)

The relief ratio may be defined as the ratio between the total relief of a basin and the longest dimension of the basin parallel to the main drainage line (Schumm 1956). Total relief is the difference in the elevation between the highest point of a watershed and the lowest point on the valley floor (Pareta and Pareta 2012) (Table 3.4).

3.2.3 Slope Instability Analysis and Sub-watersheds Wise and Prioritization Using Compound Ranking Method

The morphometric parameters i.e. bifurcation ratio (Rb), drainage density (Dd), stream frequency (Fs), texture ratio (Rt), drainage texture (Td), length of overland flow (Lof), compactness coefficient (Cc), basin shape/shape factor (Sb), form factor (Ff), circularity ratio (Rc), elongation ratio (Re) were termed as erosion risk assessment parameters (Biswas et al. 1999). In this study, two extra parameters were added for prioritization of sub-watersheds i.e. constant of channel maintenance (Ccm) and relief ratio (Rr). The linear aspects like bifurcation ratio and length of overland flow, areal aspects like drainage density, stream frequency, texture ratio, drainage texture and relief aspects like relief ratio have a direct correlation with erodibility that means higher the value, more is the erodibility or simply instability. Thus for prioritization of 18 sub-watersheds the highest value was rated as 1, second highest value was rated as 2 and so on, and the lowest value was rated as 18 (Table 3.5). On the other hand, the areal aspects such as compactness coefficient, constant of channel maintenance, basin shape, form factor, circularity ratio, elongation ratio have an inverse relationship with erodibility, lower the value, more is the erodibility (Nookaratnam et al. 2005). Hence, the lowest value was rated as 1, next lower value was rated as 2 and so on, and the highest value was rated as 18 (Table 3.5). After assigning the rank to every single parameter in relation to slope instability, the ranking values of each sub-watershed were summed up for each of the 18 sub-watersheds to arrive at compound score. Based on the compound score, the sub-watersheds having least value was assigned highest priority, next higher value was assigned second highest priority and so on.

3.3 Result and Discussion

3.3.1 Linear Aspects

The details results of linear aspects like weighted mean bifurcatin ratio (Rbwm) and length of overland flow (Lof) was shown in Table 3.6. The values of Rbwm was used to classify the watershed into five categories i.e. very low, low, moderate, high and very high. SW 1, SW 2, SW 3, SW 5, and SW 7 dominated by very high Rbwm value ranging from 3.78 to 4.10 due to low permeability and the presence of joints/fractures observed in the exposed rocks helping first order streams while SW 18 registered with very low Rbwm of 1.60 indicating less disturbed watersheds (Fig. 3.2a). Low value of Lof indicates the basin was structurally complex and vice versa (Table 3.2). The calculated values of Lof are classified into five classes i.e. very high, high, moderate, low and very low. Very high (0.262–0.857) and very low (0.069–0.084) Lof were found SW 18 and SW 4, SW 5, SW 8, SW 9 respectively (Fig. 3.2b).

Table 3.5 Ranking of 18 sub-watersheds in relation to slope instability based on morphometric parameters and 18 point scale

Morphometric parameters	Rank at 18 point scale	Logic applied
Bifurcation ratio (Rb)	1 rank to sub-watershed which has the highest bifurcation ratio	Higher the bifurcation ratio value greater the slope instability
Length of overland Flow (Lof)	1 rank at the highest length of overland flow	Higher the length of overland value, the sub-watershed become more instable
Drainage density (Dd)	1 rank at the sub-watershed having highest drainage density	The sub-watershed become more instable where the drainage density value is maximum
Stream frequency (Fs)	1 rank at the highest stream frequency	Higher stream frequency value leads more instability
Texture ratio (Rt)	1 rank at the maximum texture ratio	Texture ratio value directly related to slope instability
Drainage texture (Td)	1 rank at the largest drainage texture	Slope become more instable where drainage texture value is higher
Compactness coefficient (Cc)	1 rank at the lowest compactness coefficient	Slope instability is negatively correlated with compactness coefficient value
Constant of channel maintenance (Ccm)	1 rank at the lowest constant of channel maintenance	Higher the constant of channel maintenance value lower the slope instability
Shape factor/basin shape (Sf)	1 rank at the lowest shape factor	Higher shape factor value represents lesser slope instability
Form factor (Ff)	1 rank at the lowest form factor	Slope instability is more in lesser form factor value area.
Circularity ratio (Rc)	1 rank at the lowest circularity ratio	Basin circularity ratio value negatively related to slope instability
Elongation ratio (Re)	1 rank at the lowest elongation ratio	Lower the value of elongation ratio indicates greater slope instability
Relief ratio (Rr)	1 rank at the highest relief ratio	Higher value indicates more slope instability

Table 3.6 Morphometric parameter wise sub-watersheds fall in different categories

Parameters	Sub-watersheds fall in different category				
	Very low	Low	Moderate	High	Very high
Drainage density	SW 14, SW 18	SW 10, SW 11, SW 12, SW 15	SW 2, SW 7, SW 16, SW 17	SW 1, SW 3, SW 6	SW 4, SW 5, SW 8, SW 9
Stream frequency	SW 14, SW 18	SW 2, SW 7, SW 10, SW 11, SW 12	SW 1, SW 13, SW 15, SW 16	SW 3, SW 6, SW 8, SW 17	SW 4, SW 5, SW 9
Bifurcation ratio	SW 18	SW 6, SW 13, SW 16, SW 17	SW 4, SW 9, SW 10, SW 12, SW 14, SW 15	SW 8, SW 11	SW 1, SW 2, SW 3, SW 5, SW 7
Length of overland flow	SW 4, SW 5, SW 8, SW 9	SW 1, SW 3, SW 6, SW 17	SW 2, SW 7, SW 10, SW 11, SW 12, SW 13, SW 15, SW 16	SW 14	SW 18
Constant of channel maintenance	SW 4, SW 5, SW 8, SW 9	SW 1, SW 3, SW 6, SW 17	SW 2, SW 7, SW 10, SW 11, SW 12, SW 15, SW 16	SW 14	SW 18
Form factor	SW 8, SW 13, SW 16, SW 18	SW 15, SW 17	SW 1, SW 5, SW 7	SW 2, SW 3, SW 4, SW 6, SW 9, SW 10, SW 12, SW 14	SW 11
Circularity ratio	SW 7, SW 13, SW 18	SW 8, SW 11, SW 15, SW 16	SW 1, SW 5, SW 9, SW 10, SW 17	SW 2, SW 3, SW 4, SW 6	SW 12, SW 14
Shape factor	SW 3, SW 11	SW 2, SW 4, SW 6, SW 9, SW 10, SW 12, SW 14	SW 1, SW 5, SW 7	SW 15, SW 17	SW 8, SW 13, SW 16, SW 18
Compactness coefficient	SW 12, SW 14	SW 2, SW 3, SW 4, SW 6	SW 1, SW 5, SW 9, SW 10, SW 12, SW 14, SW 17	SW 8, SW 11, SW 15, SW 16	SW 7, SW 13, SW 18
Relief ratio	SW 18	SW 1, SW 7, SW 14, SW 15	SW 6, SW 13, SW 16, SW 17	SW 2, SW 5, SW 8, SW 9, SW 11, SW 12	SW 3, SW 4, SW 10
	Very coarse	Coarse	Moderate	Fine	Very fine
Texture ratio	SW 14, SW 18	–	SW 7, SW 10, SW 11, SW 12	SW 2, SW 13, SW 15, SW 16	SW 1, SW 3, SW 4, SW 5, SW 6, SW 8, SW 9, SW 17
Drainage texture	SW 18	SW 14	SW 7, SW 10, SW 11, SW 12	SW 2, SW 13, SW 16, SW 17	SW 1, SW 3, SW 4, SW 5, SW 6, SW 8, SW 9, SW 15
	More elongated	Elongated	Less elongated	Oval	Circular
Elongation ratio	SW 8, SW 13	SW 1, SW 2, SW 4, SW 5, SW 6, SW 7, SW 9, SW 10, SW 15, SW 16, SW 17, SW 18	SW 3, SW 12	SW 11	–

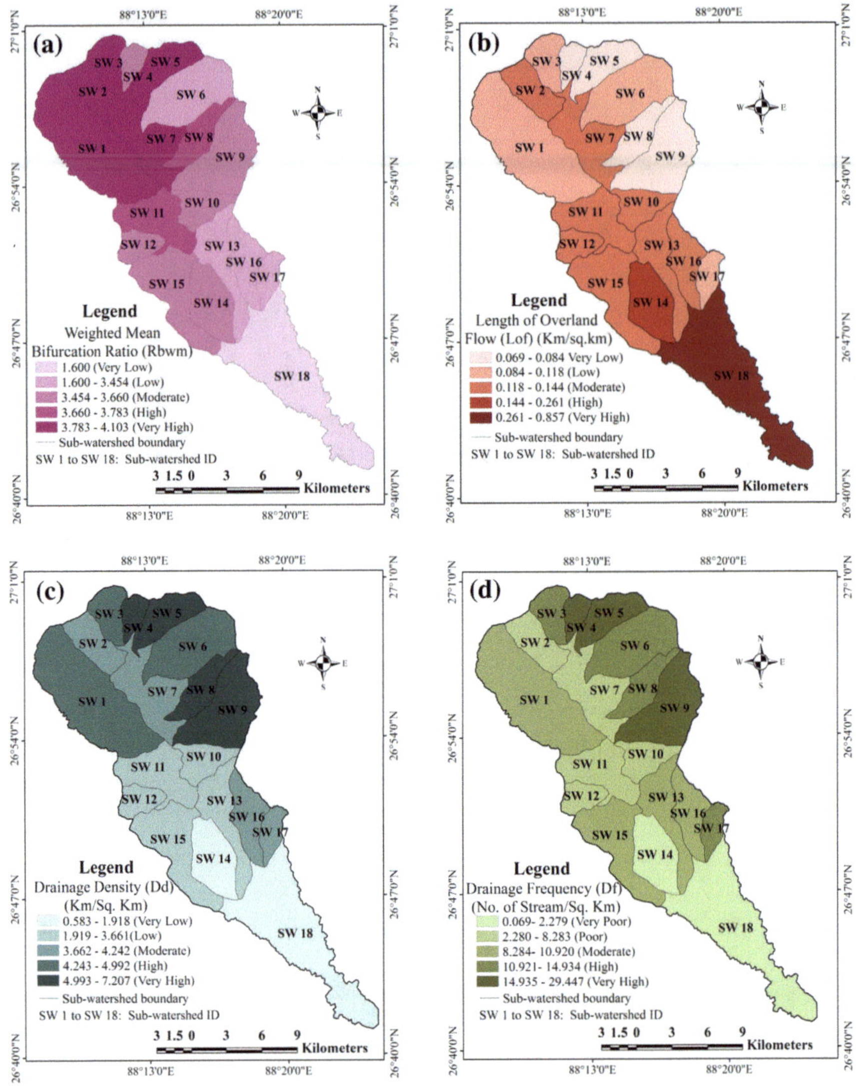

Fig. 3.2 Sub-watershed wise data layers of **a** weighted mean bifurcation ratio, **b** length of overland flow, **c** drainage density and **d** drainage frequency

3.3.2 Areal Aspects

The results of areal aspects like drainage density (Dd), drainage frequency (Fs), texture ratio (Rt), drainage texture (Td), compactness coefficient (Cc), constant of channel maintenance (Ccm), shape factor (Sf), form factor (Ff), circularity ratio (Rc) and elongation ratio (Re) were given in Table 3.6. Dd in all the sub-watersheds ranges from 0.58 to 7.21 km/km^2 (Table 3.2). The low values of Dd are associated with regions having highly permeable sub-soil materials under dense vegetation cover and low relief while high values of Dd indicate regions of weak or impermeable subsurface materials, sparse vegetation and mountainous relief. In the Balason river basin very low Dd were found in SW 14 and SW 18 (0.583–1.918) and on the other hand very high Dd regions were found in SW 4, SW 5, SW 8 and SW 9 (4.993–7.207) because of impermeable subsurface materials and high relief (Fig. 3.2c). SW 4, SW 5 and SW 9 (14.935–29.447) have very high Fs because of the presence of ridges on both sides of the valley, high relief resistant/low conducting sub surface material and indicate large number of stream availability, high runoff value, low infiltration capacity while, very poor was found in SW 14, SW 18 (0.069–2.279) indicating low relief, highly permeable subsurface material, low runoff value and small number of stream availability (Table 3.2 and Fig. 3.2d). The Rt were calculated for the 18 sub-watersheds (Table 3.2) and classified into five classes i.e. very coarse, coarse, moderate, fine and very fine (Fig. 3.3a).Very coarse texture ratio was found in SW 14 and SW 18 (<2) indicating permeable subsurface while very fine was in SW 1, SW 3, SW 4, SW 5, SW 6 and SW 9 (>8) representing impermeable subsurface. In the same way, the Td was very coarse in SW 18(<2) reveals permeable surface while very fine in SW 1, SW 3, SW 4, SW 5, SW 6, SW 8, SW 9 and SW 15 (>8) indicating impermeable subsurface material (Table 3.2 and Fig. 3.3b).

Lower value of Cc reveals less erosion while higher value indicates high erosion (Table 3.3). SW 12 and SW 14 were associated with very low Cc (1.198–1.203), on the other hand SW 7, SW 13, SW 18 were depicted with high Cc (1.622–2.156) (Fig. 3.3c). The lower value of Ccm indicates that the area is under the influence of structural disturbances having high runoff and low permeability. In this study, the value of Ccm was very high in SW 18 (0.522–1.714) and very low in SW 4, SW 5, SW 8 and SW 9 (0.139–0.167) (Table 3.2 and Fig. 3.3d). SW 3, SW 11 fall in the category of very low Sf (1.780–2.312) indicates peaked flood discharge while SW 8, SW 13, SW 16 and SW 18 have the very high shape factor (4.015–5.804) reveals weaker flood discharge periods (Table 3.3 and Fig. 3.4a). The watershed with high Ff has high peak flows of shorter duration and vice versa (Table 3.4). In present study, the Basin was classified into five parts based on form factors i.e. very high, high, moderate, low and very low. Value of form factor obtained very low for watershed SW 8, SW 13, SW 16 and SW 18 (0.172–0.202) while very high for watershed SW 11 (0.434–0.562) (Fig. 3.4b).

Based on Rc value, the Balason River basin was classified into five category i.e. very high, high, moderate, low and very low. If the value was low then the discharge will be slow as compared to the others and so the possibility of erosion will be less.

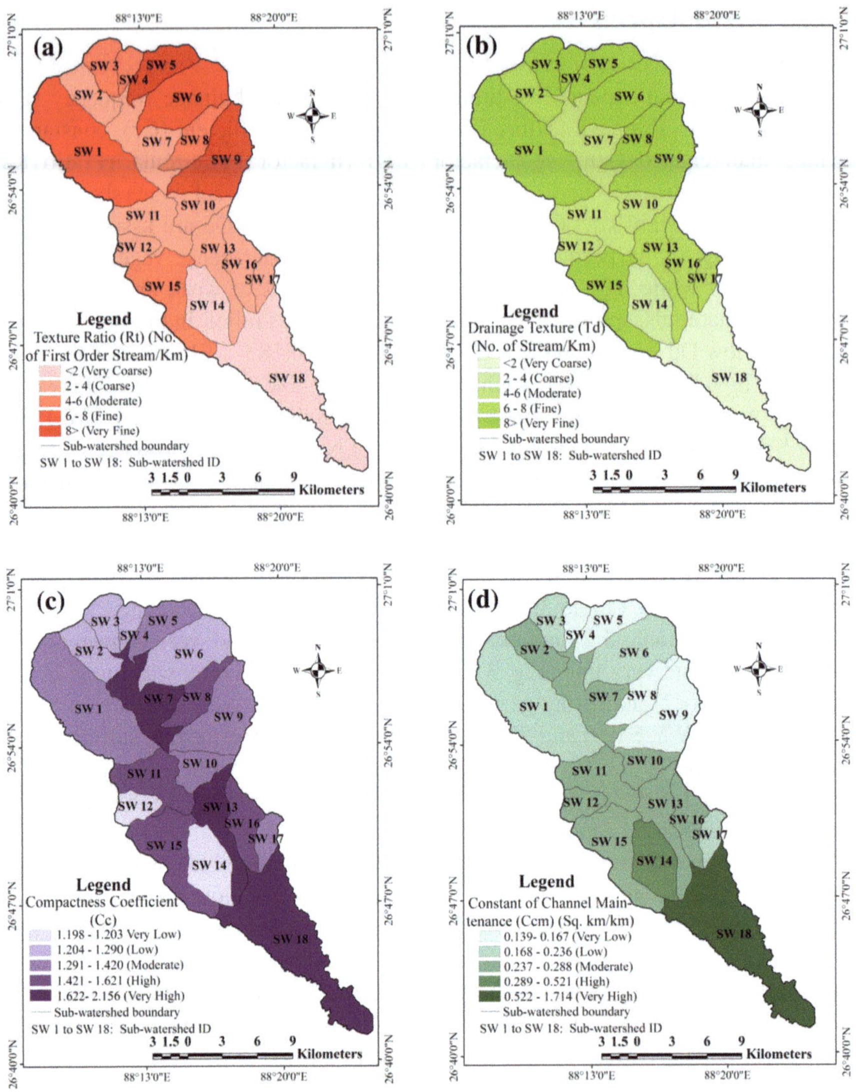

Fig. 3.3 Sub-watershed wise data layers of the Balason river basin **a** texture ratio, **b** drainage texture, **c** compactness coefficient and **d** constant of channel maintenance

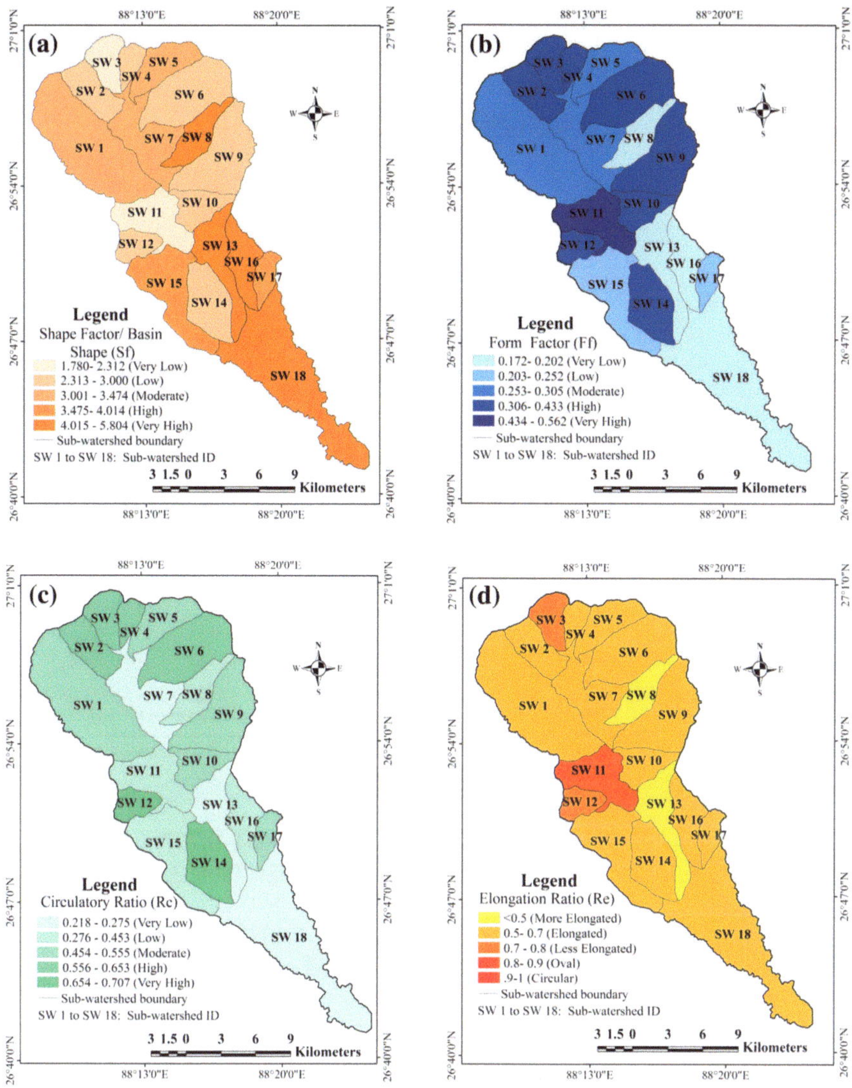

Fig. 3.4 Sub-watershed wise data layers of the Balason river basin **a** shape factor, **b** form factor, **c** circulatory ratio and **d** elongation ratio

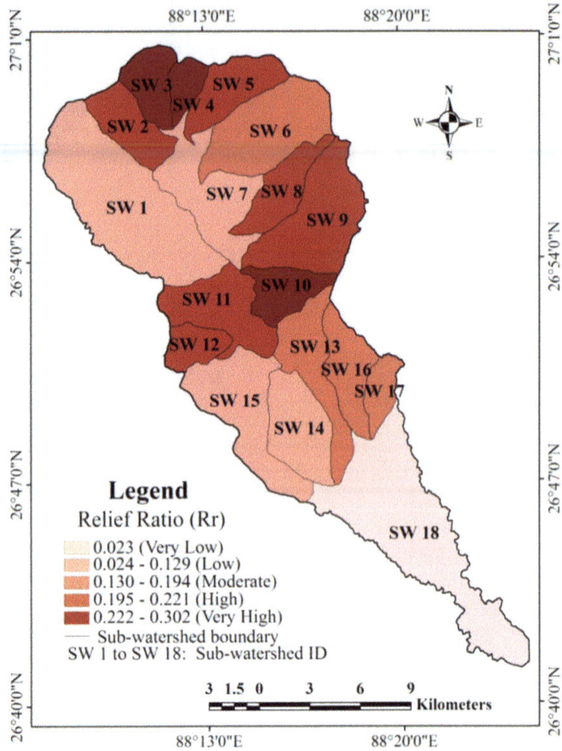

Fig. 3.5 Sub-watershed wise data layer of relief ratio of the Balason river basin

Very high value of Rc was found in SW 12 and SW 14 (0.654–0.707) while very low value observed in SW 7, SW 13 and SW 18 (0.218–0.275) (Table 3.4 and Fig. 3.4c). The oval shape sub-watershed was found in SW 11(0.8–0.9) and more elongated shape in SW 8 and SW 13 (<0.5) (Table 3.4 and Fig. 3.4d).

3.3.3 *Relief Aspects*

It measures the overall steepness of the watershed. Low values of Rr indicates lower slope and minimum intensity of erosion process while high value reveals high slope, hilly region and maximum intensity of erosion process (Table 3.3). The Rr for watersheds varies from 0.023–0.302. Very low Rr was found in SW 18 and very high Rr for SW 3, SW 4 and SW 10 (Table 3.6 and Fig. 3.5).

Table 3.7 Prioritized zones of Balason river basin

Prioritized zone	Sub-basins fall in this category
Very high	SW 4, SW 5, SW 8
High	SW 1, SW 3, SW 9
Moderate	SW 2, SW 6, SW 15, SW 16, SW 17
Low	SW 7, SW 10, SW 11, SW 13
Very low	SW 12, SW 14, SW 18

3.3.4 Prioritization of Sub-watersheds

On the basis of morphometric analysis, SW 4, SW 5 and SW 8 fall in the high priority zone while SW 12, SW 14 and SW 18 fall in the very low category (Table 3.7 and Fig. 3.6a). Sub-watershed wise highest priority obtained by SW 5 followed by SW 8, SW 4, SW 3, SW 9, SW 1, SW 6, SW 16, SW 2, SW 17, SW 15, SW 13, SW 7, SW 10, SW 11, SW 12, SW 14, and SW 18 (Table 3.8 and Fig. 3.6b).

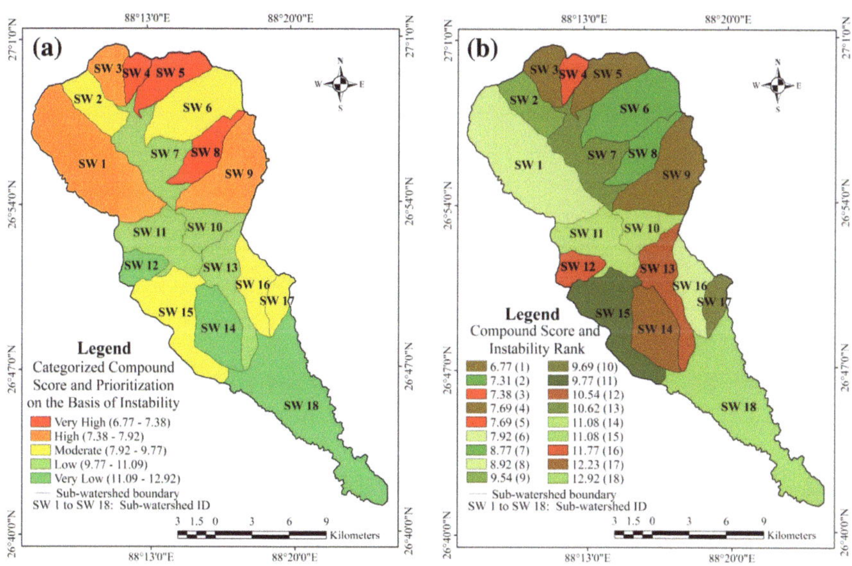

Fig. 3.6 Sub-watershed prioritization, **a** categorized compound score and prioritization, **b** individual sub-watershed prioritization

Table 3.8 Instability rank of Balason river basin based on composite ranking coefficient value

Watershed Id.	Rbwm	Dd	Fs	Td	Rt	Lof	Rr	Cc	Ccm	Sb	Ff	Rc	Re	Composite score	Compound score	Instability rank
SW 1	4.016 (2)	4.887 (7)	0.92 (8)	16.264 (3)	12.143 (3)	0.102 (12)	0.129 (14)	1.405 (10)	0.205 (7)	3.278 (10)	0.305 (9)	0.514 (9)	0.623 (9)	103	7.92	6
SW 2	4.012 (3)	3.982 (9)	8.283 (12)	6.647 (11)	4.955 (10)	0.126 (10)	0.209 (7)	1.290 (6)	0.251 (9)	2.830 (4)	0.353 (15)	0.609 (13)	0.671 (15)	124	9.54	9
SW 3	4.103 (1)	4.992 (5)	4.804 (5)	10.923 (6)	8.245 (6)	0.100 (14)	0.248 (3)	1.246 (3)	0.200 (5)	2.312 (2)	0.433 (17)	0.653 (16)	0.742 (17)	100	7.69	4
SW 4	3.532 (13)	6.538 (2)	3.023 (2)	12.335 (5)	8.851 (5)	0.076 (17)	0.302 (1)	1.265 (4)	0.153 (2)	2.969 (8)	0.337 (11)	0.634 (15)	0.655 (11)	96	7.38	3
SW 5	4.002 (4)	7.207 (1)	9.447 (1)	23.158 (1)	16.958 (1)	0.069 (18)	0.212 (15)	1.392 (8)	0.139 (1)	3.474 (12)	0.288 (7)	0.523 (11)	0.605 (7)	88	6.77	1
SW 6	3.454 (14)	4.94 (6)	2.88 (6)	15.432 (4)	10.923 (4)	0.101 (13)	0.194 (10)	1.276 (5)	0.202 (6)	2.935 (6)	0.341 (13)	0.623 (14)	0.659 (13)	114	8.77	7
SW 7	3.969 (5)	3.815 (11)	7.61 (14)	4.957 (16)	3.678 (14)	0.131 (8)	0.129 (15)	2.156 (18)	0.262 (11)	3.433 (11)	0.291 (8)	0.218 (1)	0.609 (8)	138	10.62	13
SW 8	3.783 (6)	5.986 (4)	4.934 (4)	9.765 (7)	7.055 (7)	0.084 (15)	0.209 (8)	1.554 (13)	0.167 (4)	5.556 (17)	0.180 (2)	0.420 (6)	0.479 (2)	95	7.31	2
SW 9	3.639 (10)	6.316 (3)	9.993 (3)	22.111 (2)	15.963 (2)	0.079 (16)	0.204 (9)	1.404 (9)	0.158 (3)	2.841 (5)	0.352 (14)	0.514 (10)	0.669 (14)	100	7.69	5
SW 10	3.539 (12)	3.494 (15)	7.467 (15)	5.009 (15)	3.578 (16)	0.143 (4)	0.257 (2)	1.420 (11)	0.286 (15)	2.949 (7)	0.339 (12)	0.502 (8)	0.657 (12)	144	11.08	14
SW 11	3.771 (7)	3.626 (14)	7.415 (16)	5.931 (13)	4.253 (13)	0.138 (5)	0.212 (6)	1.608 (14)	0.276 (14)	1.780 (1)	0.562 (18)	0.392 (5)	0.846 (18)	144	11.08	15
SW 12	3.636 (11)	3.646 (13)	7.672 (13)	5.047 (14)	3.678 (15)	0.137 (6)	0.221 (4)	1.198 (1)	0.274 (13)	2.552 (13)	0.392 (16)	0.707 (18)	0.706 (16)	153	11.77	16
SW 13	3.432 (16)	3.478 (16)	9.802 (10)	6.271 (12)	4.421 (12)	0.144 (3)	0.165 (13)	1.919 (16)	0.288 (16)	5.804 (18)	0.172 (1)	0.275 (3)	0.468 (1)	137	10.54	12

(continued)

Table 3.8 (continued)

Watershed Id.	Rbwm	Dd	Fs	Td	Rt	Lof	Rr	Cc	Ccm	Sb	Ff	Rc	Re	Composite score	Compound score	Instability rank
SW 14	3.66 (8)	1.918 (17)	2.279 (17)	2.366 (17)	1.72 (17)	0.261 (2)	0.102 (16)	1.203 (2)	0.521 (17)	2.999 (9)	0.333 (10)	0.701 (17)	0.652 (10)	159	12.23	17
SW 15	3.647 (9)	3.661 (12)	9.358 (11)	8.611 (8)	6.175 (8)	0.137 (7)	0.099 (17)	1.621 (15)	0.273 (12)	4.014 (14)	0.249 (5)	0.386 (4)	0.563 (5)	127	9.77	11
SW 16	3.011 (17)	3.852 (10)	10.786 (9)	7.665 (9)	5.127 (9)	0.130 (9)	0.192 (11)	1.496 (12)	0.260 (10)	4.944 (15)	0.202 (4)	0.453 (7)	0.507 (4)	116	8.92	8
SW 17	3.445 (15)	4.242 (8)	12.315 (7)	6.721 (10)	4.777 (11)	0.118 (11)	0.174 (12)	1.351 (7)	0.236 (8)	3.966 (13)	0.252 (6)	0.555 (12)	0.567 (6)	126	9.69	10
SW 18	1.6 (18)	0.583 (18)	0.069 (18)	0.069 (18)	0 (18)	0.857 (1)	0.023 (18)	2.147 (17)	1.714 (18)	5.026 (16)	0.199 (3)	0.220 (2)	0.503 (3)	168	12.92	18

3.4 Conclusion

Watershed prioritization is considered as one of the most important aspects of planning, development and management programmes of a geomorphic unit over the earth surface. The assessment of morphometric parameters using GIS helped to understand the existing geomorphic process operating within the framework of drainage basin which also helps for prioritization of watersheds. Balason is a seventh order river basin of Darjeeling Himalaya. 13 morphometric parameters were used for hierarchical arrangement of sub-watersheds in the Balason river basin. It can realistically be assumed that watersheds which attained first priority likely to have a high level of slope instability problems and should be considered for immediate conservation measures. Therefore, immediate attention towards the management of land, water and soil are required in the sub-watersheds having low compound score and low rank. The sub-watersheds experienced with low compound scores are to be brought under land use planning, conservation of biodiversity, and development.

References

Ali U, Ali SA (2014) Analysis of drainage morphometry and watershed prioritization of Romushi-Sasar catchment, Kashmir Valley, India using remote sensing and GIS technology. Int J Adv Res 2:5–23

Amani M, Safaviyan A (2015) Sub-basins prioritization using morphometric analysis remote sensing technique and GIS Golestan—Iran. Int Lett Nat Sci 5–6 (online)

Biswas S, Sudhakar S Desai, VR (1999) Prioritization of sub-watersheds based on morphometric analysis of drainage basin: a remote sensing and GIS approach. J Indian Soc Remote Sens 27:155–166

Chirala U, Gurram MK, Nookaratnam K (2015) Mapping of soil erosion zones of Meghadrigedda catchment, Visakhapatnam, India for conservation—a geospatial approach. Int J Geosci 6:326–338

Chopra R, Dhiman RD, Sharma P (2005) Morphometric analysis of subwatersheds in Gurdaspur district, Punjab using remote sensing and GIS techniques. J Indian Soc Remote Sens 33:531–539

Chorley RJ (1969) The drainage basin as the fundamental geomorphic unit. In: Chorley RJ (ed) Water, earth and man. Methuen, London, pp 77–100

Gravelius H (1914) Grundrifi der gesamtangewcisserkunde; Band I: Flufikunde (compendium of hydrology, vol I Rivers, in German), Goschen, Berlin, Germany

Gupta M, Srivastava PK (2010) Integrating GIS and remote sensing for identification of ground water potential zones in the hilly terrain of Pavagarh, Gujarat, India. Water Int 35:233–245

Horton RE (1932) Drainage basin characteristics. Am Geophys Union 13:350–361

Horton RE (1945) Erosional development of streams and their drainage basins, a hydrophysical approach to quantitative morphology. Geol Soc Am Bull 56:275–370

Javed A, Khanday MY, Ahmed R (2009) Prioritization of sub-watersheds based on morphometric and land use analysis using RS and GIS technology. J Indian Soc Remote Sens 37:261–274

Javed A, Khanday MY, Rais S (2011) Watersheds prioritization using morphometric and land use/land cover parameters: a remote sensing and GIS based approach. J Geol Soc India 78:63–75

Magesh NS, Jitheshlal KV, Chandrasekar N, Jini KV (2012) GIS based morphometric evaluation of Chimmini and Mupily watersheds, parts of Western Ghats, Thrissur District, Kerala, India. Earth Sci Inform 5(2):111–121

Melton MA (1957) An analysis of the relations among the elements of climate, surface properties and geomorphology. Technical report 11, Department of Geology, Columbia University, New York

Melton MA (1958) Correlations structure of morphometric properties of drainage systems and their controlling agents. J Geol 66, 442–460

Miller VC (1953) A quantitative geomorphic study of drainage basin characteristics in the clinch mountain area. Virginia and Tennessee, Department of Geology, Columbia University, Contract N6 ONR 271-30, technical report 3, pp 1–30

Nautiyal MD (1994) Morphometric analysis of a drainage basin, district Dehradun, Uttar Pradesh. J Indian Soc Remote Sens 22(4):251–261

Nookaratnam K, Srivastava YK, VenkateshwaraRao V, Amminedu E, Murthy KSR (2005) Check dam positioning by prioritization of microwatersheds using SYI model and zorphometric analysis—remote sensing and GIS perspective. J Indian Soc Remote Sens 33:25–38

Panda RK, Nagarajan R (2013) Prioritization of sub-basins of Ghataprabha river using morphometric parameters. Int J Geomat Geosci 3:605–617

Panhalkar SS, Pawar CT (2011) Watershed development prioritization by applying WERM model and GIS technique in Vedganga basin (India). ARPN Am J Agric Biol Sci 38–44

Pareta K, Pareta U (2012) Quantitative morphometric analysis of a watershed of Yamuna basin, India using ASTER (DEM) data and GIS. Int J Geomat Geosci 2:248–269

Patel DP, Gajjar CA, Srivastava PK (2013) Prioritization of Malesari mini-watersheds through morphometric analysis: a remote sensing and GIS perspective. Environ Earth Sci 69:2643–2656

Patil VSP, Mali SP (2013) Watershed characterization and prioritization of Tulsi sub-watershed: a geospatial approach. Int J Innov Res Sci Eng Technol 2:2182–2189

Parveen R, Kumar U, Singh VK (2012) Geomorphic characterization of upper south Koel basin, Jharkhand: a remote sensing and GIS approach. J Water Resour Protect 4:1042–1050

Rekha VB, George AV, Rita M (2011) Morphometric analysis and microwatersheds prioritization of Peruvanthanam subwatershed, the Manimala river basin, Kerala, South India. Environ Res Eng Manag 3:6–14

Schumm SA (1956) Evaluation of drainage system and slopes in Badlands at Perth Amboy, New Jersey. Geol Soc Am Bull 67:597–646. https://doi.org/10.1130/0016-7606(1956)67%5b597:EODSAS%5d2.0.CO;2)

Sharma H, Padmaja G (1982) Quantitative fluvial characteristics of streams of the Mej basin (Rajasthan). In: Sharma HS (ed) Perspectives in geomorphology, quantitative fluvial geomorphology, concept, New Delhi, pp 143–190

Shrimali SS, Agarwal SP, Samra JS (2011) Prioritization erosion prone areas in hills using remote sensing and GIS: a case study of Sukhna lake catchment, North India. J Appl Geol 3:54–60

Singh S (2010) Geomorphology. Prayag Pustak Bhawan, Allahabad, pp 371–374

Singh S, Ojha SS, Agnihotri SP (1984) On the regional variations in bifurcations ratios. In: Mukhopadhayay SC (ed) Geographical mosaic, Calcutta University, pp 203–210

Smith KG (1950) Standards of grading texture of erosional topography. Am Jour Sci 248:655–668

Srivastava PK, Mukherjee S, Gupta S, Singh SK (2011) Characterizing monsoonal variation on water quality index of river Mahi in India using geographical information system. Water Qual Expo Health 2:193–203

Strahler AN (1952) Dynamic basis of geomorphology. Geol Soc Am Bull 63:923–938

Strahler AN (1957) Quantitative analysis of watershed geomorphology. Trans Am Geophys Union 38:913–920

Strahler AN (1964) Quantitative geomorphology of drainage basins and channel networks. In: Chow VT (ed) Handbook of applied hydrology, section 4-11. McGraw Hill Book Company, New York

Tolessa GA, Rao PJ (2013) Watershed development prioritization of Tandava river basin, Andhra Pradesh, India—GIS approach. Int J Eng Sci Inven 2:12–20

Vandana M (2012) Morphometric analysis and watershed prioritization: a case study of Kabani river basin, Wayanad district, Kerala, India. Indian J Geomarine Sci 42:211–222

Chapter 4
Geomorphic Diversity and Landslide Susceptibility: A Multi-criteria Evaluation Approach

4.1 Introduction

Landslides are one of the real normal risks which caused roughly 1000 individuals deaths per year worldwide and around 4 billion US property damages (Lee and Pradhan 2007). The Himalayan mountain regions, the Meghalaya plateau and the Western Ghats are the most significant landslide prone areas in India. The Balason river basin of Darjeeling Himalaya is not an exception of these phenomena due to its fragile geo-environmental conditions and operating geomorphic process. Every year a large number of populations, property and natural environment were affected by landslides. So as to limit the impact of landslide damages, it is important to foresee before its event. Hence, the landslide susceptibility map is turn out to be exceptionally valuable approach to assess, manage and mitigate the landslide hazard for a region (Kienholz 1978; Spiker and Gori 2000; Chau et al. 2004). The drainage basin morphometry plays an important role in quantifying the relief, drainage and hydrological condition which governs the whole landscape of a river basin. The morphometric parameters are intimately associated with landslide activities. These parameters portray the surface chemistry and play vital role in slope instability. Several authors calculated morphometric parameters for terrain analysis using remote sensing (RS) and geographic information system (GIS) in different parts of the world (Rastogi and Sharma 1976; Nautiyal 1994; Magesh et al. 2012). Some researchers tried to find out the ground water potential zones using morphometric parameters (Gajbhiye et al. 2014; Gopal and Saha 2015). Sub-basin prioritization (Thakkar and Dhiman 2007; Amani and Safaviyan 2015), surface water availability (Rai et al. 2014; Khatun and Pal 2017), recharge estimation (Avinash et al. 2014; Nidhi et al. 2016) were done using morphometric parameters. Several studies were established to show the relationship between slope instability or simply landslide occurrences and

S. Mandal and S. Mondal, *Geoinformatics and Modelling of Landslide Susceptibility and Risk*, Environmental Science and Engineering,
https://doi.org/10.1007/978-3-030-10495-5_4

drainage morphometric parameters (Jeganathan and Chauniyal 2000; Ghosh 2015; Kumar and Garbyal 2016). In this study, total 14 morphometric parameters were calculated and spatial data layers were made accordingly. With the help of these parameters, three geomorphic diversity models i.e. drainage diversity (DD) model, relief diversity (RD) model and morphometric diversity (MD) model were applied to prepare landslide susceptibility maps. Also try to see how much each model predicts future landslide events. Basically high diversified areas are associated with high landslides prone areas and vice versa. The other existing landslide susceptibility methods like frequency ratio technique (Lee and Talib 2005; Lee and Sambath 2006), logistic regression method (Dai and Lee 2002; Lee and Sambath 2006; Ghosh et al. 2011), information value (Info Val) method (Anbalagan 1992), landslide normal risk factor method (Yin and Yin 1988), discriminant analysis method (Gupta and Joshi 1990), artificial neural network (ANN) method (Ardizzone et al. 2002; Choi et al. 2011), neuro-fuzzy method (Pradhan 2010), support vector machine (Xu et al. 2012), decision tree method (Nefeslioglu et al. 2010), index of entropy (Devkota et al. 2013a, b), bayesian probability (Chung and Fabbri 1999; Song et al. 2012) and fractal theory (Majtan et al. 2002) etc. were prepared in combination of several relief, drainage, geological, anthropological, protective and triggering factors. But in this study, the three aforementioned diversity models helps us to find out the role of drainage, relief and morphometric parameters in landslide susceptibility assessment individually. The DD model was calibrated using drainage density (Dd), drainage frequency (Df), bifurcation ratio (Rb), relative relief (Rr), drainage intensity (Din), drainage texture (Dt), junction frequency (Jf), infiltration number (In) and ruggedness index (Ri) whereas RD model was constructed using slope (S), Rr, dissection index (Di), elevation (E), Ri, Rb, relief ratio (Rra) and hypsometric integral/relative height (Hi). Finally, Dd, Df, Din, Di, Dt, Jf, In, Rb, Rr, Ri, E, S and Hi were used for MD model.

4.2 Materials and Methods

In the present study, total 14 parameters such as drainage density, drainage frequency, bifurcation ratio, relative relief, drainage intensity, drainage texture, junction frequency, infiltration number, ruggedness index, dissection index, elevation, slope, relief ratio and hypsometric integral/relative height were considered for the landslide susceptibility assessment and prediction which were prepared from SRTM DEM, Toposheets and Google earth imagery (Table 4.1 and Fig. 4.1). Various classification methods exist, such as quantile, equal interval, natural break, and standard deviation methods, to classify the aforesaid data layers (Pradhan and Lee 2010a, b). If the data distribution has a positive or negative skewness, the quantile or natural break distribution classifiers could be chosen (Akgun et al. 2012). As suggested by many researchers, natural break method was applied to determine different subclasses (Poli and Sterlacchini 2007; Althuwaynee et al. 2014; Mahalingam et al. 2016; Shrestha et al. 2017; Mandal and Mandal 2017) for this study.

Table 4.1 Dataset used to assess geomorphic diversity in the Balason river basin

Purpose	Data used	Data source
Preparation of landslide inventory map	Google earth historical imageries (2000–2016), Landsat 7 ETM+ imageries (2000–2005), Landsat 4–5 TM imageries (2006–2007, 2009–2011), LISS III imageries (2008, 2012–2013), Landsat 8 OLI (2014–2015), high resolution Sentinel-2 imagery (2016), Arc Map high resolution world imagery and intensive field survey data (2015–2016)	https://earthexplorer.usgs.gov/, https://glovis.usgs.gov, http://bhuvan.nrsc.gov.in/, Google earth, Arc Map 10.1 world imagery and Field investigation using GPS
Preparation of relative relief, dissection index, elevation, slope, hypsometric integral data layers	SRTM DEM (30 m spatial resolution) (2014)	http://earthexplorer.usgs.gov
Thematic data layers of drainage density, drainage frequency, bifurcation ratio, drainage intensity, drainage texture and junction frequency, infiltration number	Toposheet (1961–62, 1965–66, 1969–71) and Google earth imagery (2015)	Survey of India (SOI), Kolkata. toposheet no. 78A/4, 78A/8, 78B/1, 78B/5 and 78B/6 and Google earth map
Data layers of ruggedness index and relief ratio	SRTM DEM (30 m spatial resolution) and Toposheet (1961–62, 1965–66, 1969–71) and Google earth imagery (2015)	http://earthexplorer.usgs.gov, Survey of India (SOI), Kolkata. toposheet no. 78A/4, 78A/8, 78B/1, 78B/5 and 78B/6 and Google earth map

4.2.1 Preparation of Landslide Inventory Map

Landslide inventory map is the preliminary step to prepare landslide susceptibility map. Several researchers incorporated aerial photographs and satellite images to identify the different locations of landslide event (Lee and Pradhan 2007; Pradhan 2010; Choi et al. 2011). In this study, Google earth historical imageries (2000–2016), Landsat 7 ETM+ imageries (2000–2005), Landsat 4–5 TM imageries (2006–2007, 2009–2011), LISS III imageries (2008, 2012–2013), Landsat 8 OLI (2014–2015), high resolution Sentinel-2 imagery (2016), Arc Map high resolution world imagery and intensive field survey data (2015–2016) were used to the mapping of several

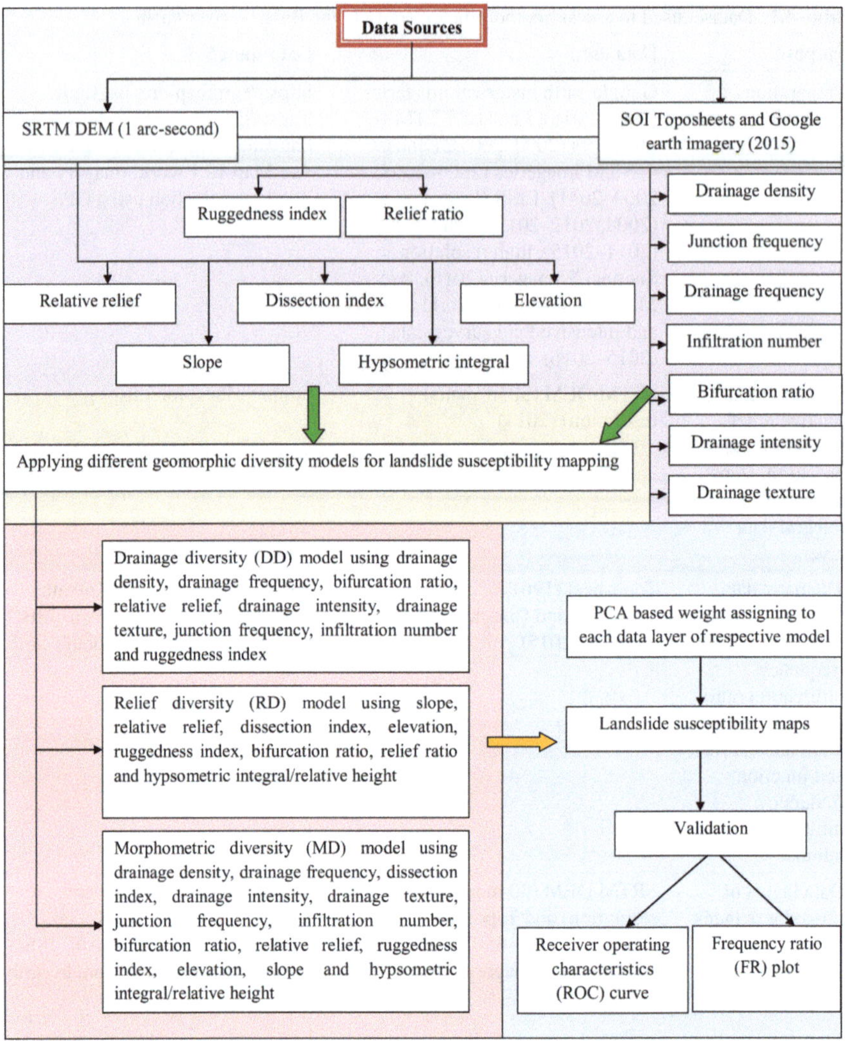

Fig. 4.1 Methodology of the study

large and small landslides. There are three major types of landslides were identified in the study area—shallow translational debris slides, deep-seated rockslides and shallow translational rockslides. A total of 295 landslide locations were mapped (see Fig. 1.1) which were randomly divided into two parts i.e. 70% of the total landslide locations were used as training data set and the remaining 30% locations were used for validation of the model. The average area of landslide was 3762.90 m^2.

4.2.2 *Morphometric Data Layers Preparation and Their Role in Slope Instability*

Total 14 parameters were calculated using recognized formula and prepared spatial data layers in Arc GIS environment (Table 4.2). The relationship between morphometric parameters and landslide occurrences were judge using frequency ratio (FR) approach (Karim et al. 2011; Mondal and Mandal 2017b).

$$FR = \frac{\{\text{N pix (Si)/N pix(Ni)}\} \times 100}{\left\{\sum \text{N pix (Si)}/\sum \text{N pix (Ni)}\right\} \times 100} \tag{4.1}$$

where, N pix (Si) is the number of pixels containing landslide in each class (i), N pix (Ni) is the total number of pixels having class (i) in the whole basin, $\sum$N pix (Si) total number of pixels in each class and $\sum$N pix (Ni) total number of pixels in the whole basin. Higher the FR value, more the chance of landslide occurrences.

4.2.2.1 Drainage Density

It is an important factor of landslide susceptibility. Drainage Density means the total stream length per unit area. It also indicates the degree of closeness of spacing of channel per unit area which determines landscape dissection and runoff potentiality of a region (Strahler 1964). Drainage density indirectly presents the erosional status of the basin and highly correlated with landslide susceptibility. Higher drainage density of an area indicates high landslide probability and lower drainage density means vice versa (Nag 1998; Sarkar and Kanungo 2004). High drainage density of an area implies high runoff, consequently low infiltration rate, whereas, low drainage density of an area implies high runoff, and consequently low drainage density of an area refers to low runoff and high infiltration. In the present study, drainage density varies from 0 to 15.10 km/km^2 and it was classified into ten different classes i.e. 0–1.31, 1.31–2.61, 2.61–3.67, 3.67–4.56, 4.56–5.39, 5.39–6.27, 6.27–7.28, 7.28–8.58, 8.58–10.60 and 10.60–15.10 (Fig. 4.2a).

4.2.2.2 Drainage Frequency

Drainage frequency is an important geomorphic concept which measures the relative spacing at drainage lines. Drainage frequency refers to the number of stream segments of all orders per unit area (Horton 1932, 1945). It helps for understanding of the drainage texture as well as the texture of the landscape of an area. The high drainage frequency indicates the fast runoff and therefore landslide susceptibility is more

Table 4.2 Formulae for calculating of different parameters used in drainage diversity (DD), relief diversity (RD) and morphometric diversity (MD) method

Parameters	Used in	Methods/formulae	References	Units	Results
Drainage density (Dd)	DD and MD method	Dd = L/A L = Total length of streams, A = Area of watershed	Horton (1945)	km/km^2	0–15.10
Drainage frequency (Df)	DD and MD method	Df = N/A N = Total number of streams, A = Area of watershed	Horton (1945)	No. of stream/km^2	0–104.30
Bifurcation ratio (Rb)	DD, RD and MD method	Rb = Nu/Nu + 1 Nu = Total number of stream segment of order 'u', Nu + 1 = Number of stream segment of next higher order	Schumm (1956)	Unit less	1.50–5.66
Relative relief (Rr)	DD, RD and MD method	Rr = Highest altitude – Lowest altitude	Smith (1935)	m/km^2	5.19–662
Drainage intensity (Din)	DD and MD method	Din = Df/Dd	Faniran (1968)	Din/km^2	0–48.17
Drainage texture (Dt)	DD and MD method	Dt = Ns/P Ns = Total number of streams, P = Perimeter of watershed	Horton (1945)	No. of stream/km	0–26.07
Junction frequency (Jf)	DD and MD method	Jf = No. of stream node/Area	–	No. of stream node/km^2	0–67.49
Infiltration number (In)	DD and MD method	In = Df × Dd	Faniran (1968)	In/km^2	0–1575.50

(continued)

Table 4.2 (continued)

Parameters	Used in	Methods/formulae	References	Units	Results
Ruggedness index (Ri)	DD, RD and MD method	Ri = (Rr × Dd)K K = constant, 5280 for mile map, 1000 for km map	Miller et al. (1990)	Ri/km^2	0–6.55
Slope (S)	RD and MD method	Using 3D analyst tool in Arc GIS 10.1 software	–	Degree	0–71.10
Dissection index (Di)	RD and MD method	Di = Rr/Ar Ar = Absolute relief	Dove (1957)	Di/km^2	0.06–0.59
Elevation (E)	RD and MD method	Elevation map was prepared from SRTM DEM	–	m	106–2600
Relief ratio (Rra)	RD method	Rra = Bh/Lb Bh = Basin relief, Lb = Basin length	Schumm (1956)	Unit less	0.02–0.30
Hypsometric integral/relative height (Hi)	RD and MD method	Hi = (Mean E-Minimum E)/(Maximum E-Minimum E)	Strahler (1964)	Unit less	0.18–0.57

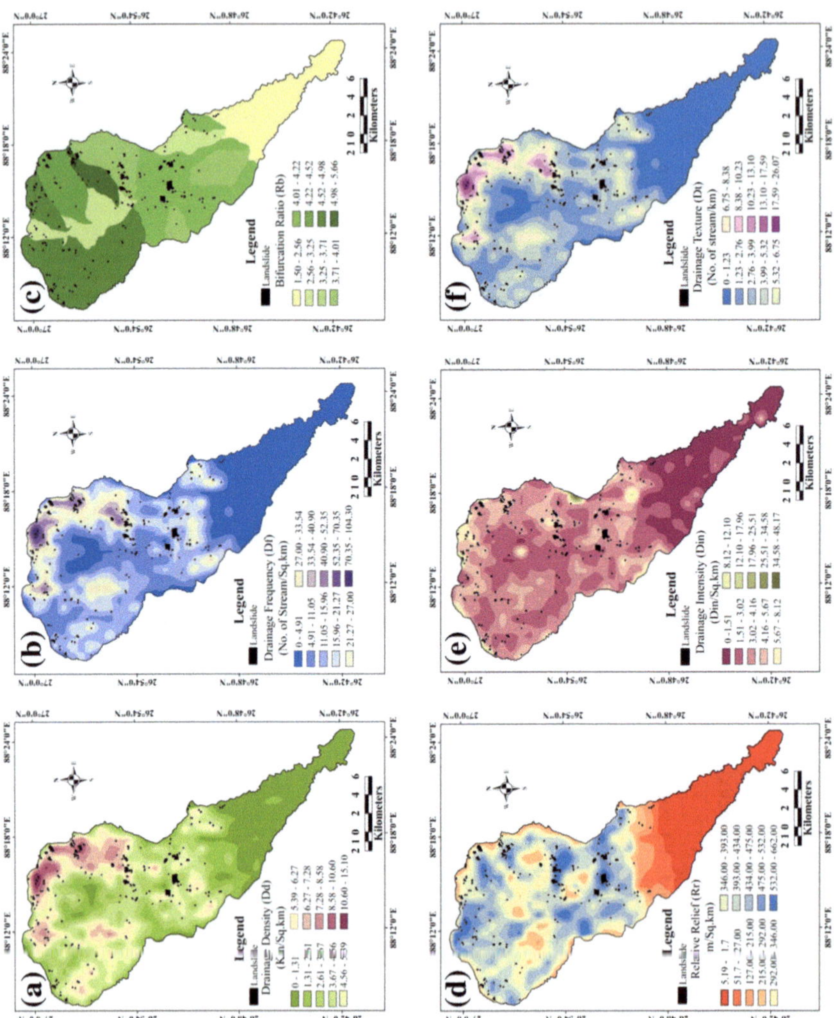

Fig. 4.2 Spatial data layers of **a** drainage density, **b** drainage frequency, **c** bifurcation ratio, **d** relative relief, **e** drainage intensity, **f** drainage texture, **g** junction frequency, **h** infiltration number, **i** ruggedness index, **j** dissection index, **k** elevation and **l** slope

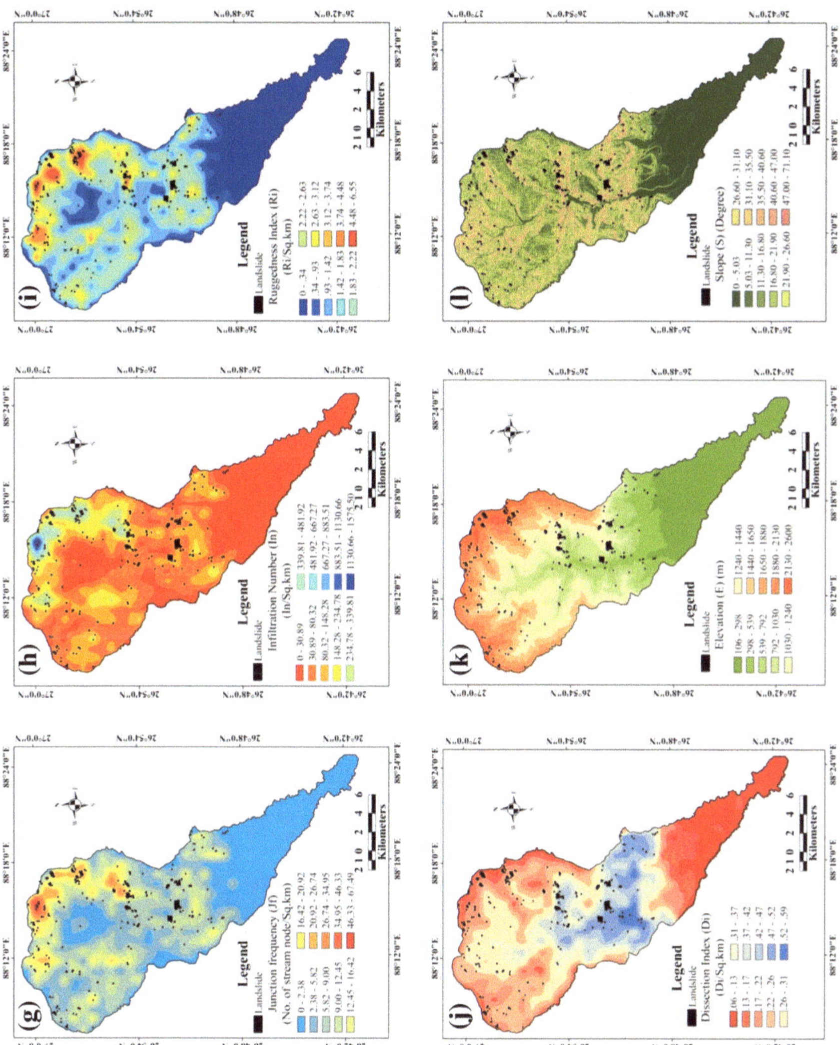

Fig. 4.2 (continued)

likely in the basin with high stream frequency (Jeganathan and Chauniyal 2000). In this study, drainage frequency was grouped into 10 classes: 0–4.91, 4.91–11.05, 11.05–15.96, 15.96–21.27, 21.27–27.00, 27.00–33.54, 33.54–40.90, 40.90–52.35, 52.35–70.35 and 70.35–104.30 no. of stream/km^2 (Fig. 4.2b).

4.2.2.3 Bifurcation Ratio

Bifurcation ratio is a dimensionless property of the drainage basin may be defined as the ratio between the total number of stream segment of an order and next higher order (Horton 1945). It is strongly controlled by drainage density, lithological characteristics, basin shape and area, and stream entrance angle etc. Higher the bifurcation ratio means structurally disturbed watersheds with high distortion in drainage pattern indicate high landslide prone area (Ghosh 2015). Bifurcation ratio of the Balason river basin ranges from 1.50 to 5.66 and it was reclassified into eight classes i.e. 1.50–2.56, 2.56–3.25, 3.25–3.71, 3.71–4.01, 4.01–4.22, 4.22–4.52, 4.52–4.98 and 4.98–5.66 (Fig. 4.2c).

4.2.2.4 Relative Relief

Relative relief is another important parameter to prepare landslide susceptibility zonation map (Mandal and Mandal 2016). It can be defined as the difference between highest altitude and lowest altitude. The relative relief is positively associated with landslide susceptibility. High relative relief enhances the chance of slope failure and vice versa (Jeganathan and Chauniyal 2000). For the present study, the relative relief value ranges from 5.19 to 662.00 m/km^2 and divided into 10 different classes i.e. 5.19–51.70, 51.70–127.00, 127.00–215.00, 215.00–292.00, 292.00–346.00, 346.00–393.00, 393.00–434.00, 434.00–475.00, 475.00–532.00 and 532.00–662.00 (Fig. 4.2d).

4.2.2.5 Drainage Intensity

Drainage intensity may be defined as the ratio of the drainage frequency to the drainage density (Faniran 1968). High value of drainage intensity indicates that drainage density and stream frequency have significant influence on the degree to which land surface of the watershed has been lowered by various denudation processes. Simply high drainage intensity areas are characterized by more slope failure events. The drainage intensity of the study area ranges from 0 to 48.17 and reclassified it into 10 categories i.e. 0–1.51, 1.51–3.02, 3.02–4.16, 4.06–5.67, 5.67–8.12, 8.12–12.10, 12.10–17.96, 17.96–25.51, 25.51–34.58 and 34.58–48.17 (Fig. 4.2e).

4.2.2.6 Drainage Texture

Drainage texture denotes the total number of stream segments of all orders per perimeter of that area (Horton 1945). Drainage texture indicates the relative spacing of drainage lines and the cumulative effect of the infiltration capacity, the amount and type of vegetation, precipitation, infiltration capacity, viz., absorption capacity of soil, underlying lithology, and relief aspect of the terrain influences the rate of surface run-off and affects the drainage texture of a basin. High textural area indicates greater susceptibility of landslide occurrences (Jeganathan and Chauniyal 2000). In this study, the value of drainage texture ranges from 0 to 26.07 no. of stream/km and it categorized into 10 classes i.e. 0–1.23, 1.23–2.76, 2.76–3.99, 3.99–5.32, 5.32–6.75, 6.75–8.38, 8.38–10.23, 10.23–13.10, 13.10–17.59 and 17.59–26.07(Fig. 4.2f).

4.2.2.7 Junction Frequency

It is the number of stream nodes per unit area (Ghosh and Saha 2015). It reflects the hydrological condition of a basin. Higher the junction frequency value, greater chance of water availability. River erosional process leads to the slope become more vulnerable. Simply, it enhances the chance of landslide occurrences (Jeganathan and Chauniyal 2000). The value of junction frequency ranges from 0 to 67.49 no. of stream node/km^2 and it grouped into 10 classes i.e. 0–2.38, 2.38–5.82, 5.82–9.00, 9.00–12.45, 12.45–16.42, 16.42–20.92, 20.92–26.74, 26.74–34.95, 34.95–46.33 and 46.33–67.49 (Fig. 4.2g).

4.2.2.8 Infiltration Number

Infiltration number of a drainage basin is the product of drainage density and stream frequency (Faniran 1968). The higher the infiltration number the lower will be the infiltration and higher will be the run-off. Therefore, high infiltration number is closely associated with slope instability (Ghosh 2015). The value of infiltration number was reclassified into 10 classes i.e. 0–30.89, 30.89–80.32, 80.32–148.28, 148.28–234.78, 234.78–339.81, 339.81–481.92, 481.92–667.27, 667.27–883.51, 883.51–1130.66 and 1130.66–1575.50 In/km^2 (Fig. 4.2h).

4.2.2.9 Ruggedness Index

Topographic ruggedness exhibit significant control in spatial occurrence of landslides and erosion pone areas and it portrays intricate relationship of lithology, structure, relief and climate. It is the product of the product of drainage density and basin relief divided by 1000 (Miller et al. 1990). The low ruggedness value of basin implies that area is less prone to soil erosion, landslide and have intrinsic structural complexity in association with relief and drainage density. In our study it ranges from 0 to

6.55 Ri/km^2 which partitioned into 10 groups i.e. 0–0.34, 0.34–0.93, 0.93–1.42, 1.42–1.83, 1.83–2.22, 2.22–2.63, 2.63–3.12, 3.12–3.74, 3.74–4.48 and 4.48–6.55 (Fig. 4.2i).

4.2.2.10 Dissection Index

Dissection Index depicts the degree of dissection or vertical erosion, and the stage of landform development in a basin watershed. Dissection index may be defined as the ratio between the relative relief and absolute relief of the basin (Dove 1957) and the value always ranges between 0.0 (complete absence of dissection and thus the dominance of flat topography) and 1 for infrequent cases such as vertical cliff topography at the sea shore, or vertical escarpment of hill slope (Farhan et al. 2015). High dissection index enhances the chance of slope failure and vice versa (Jeganathan and Chauniyal 2000; Ghosh 2015). In this study, ten classes of dissection index were made: 0.06–0.13, 13–0.17, 0.17–0.22, 0.22–0.26, 0.26–0.31, 0.31–0.37, 0.37–0.42, 0.42–0.47, 0.47–0.52 and 0.52–0.59 Di/km^2 (Fig. 4.2j).

4.2.2.11 Elevation

Elevation is primary causative factor which affect the slope material by weathering process. Elevation has direct influence on all others topographic aspects, slope, soil layer arrangement etc. as the elevation increases the slope angle also increases which directly makes the land surface more landslide susceptible (Devkota et al. 2013a, b; Umar et al. 2014). In the present study, the elevation value ranges from 106 to 2600 m and divided into 10 classes i.e. 106–298, 298–539, 539–792, 792–1030, 1030–1240, 1240–1440, 1440–1650, 1650–1880, 1880–2130 and 2130–2600 (Fig. 4.2k).

4.2.2.12 Slope

Slope is a key factor which strongly influences the landslide susceptibility of a region. A positive correlation exists between landslides and slope which indicate that areas with steep slopes are more prone to landslide occurrences (Foumelis et al. 2004; Mandal and Mandal 2018). As the slope angle increases, the tendency of movement increases and shear stress in soil and other un-consolidated material generally increases. Moderate slopes are expected to have a low frequency of landslide (Ayalew et al. 2005). Sl of the basin ranges from 0° to 71.10° and was reclassified into ten classes i.e. 0°–5.03°, 5.03°–11.30°, 11.30°–16.80°, 16.80°–21.90°, 21.90°–26.60°, 26.60°–31.10°, 31.10°–35.50°, 35.50°–40.60°, 40.60°–47.00° and 47.00°–71.10° (Fig. 4.2l).

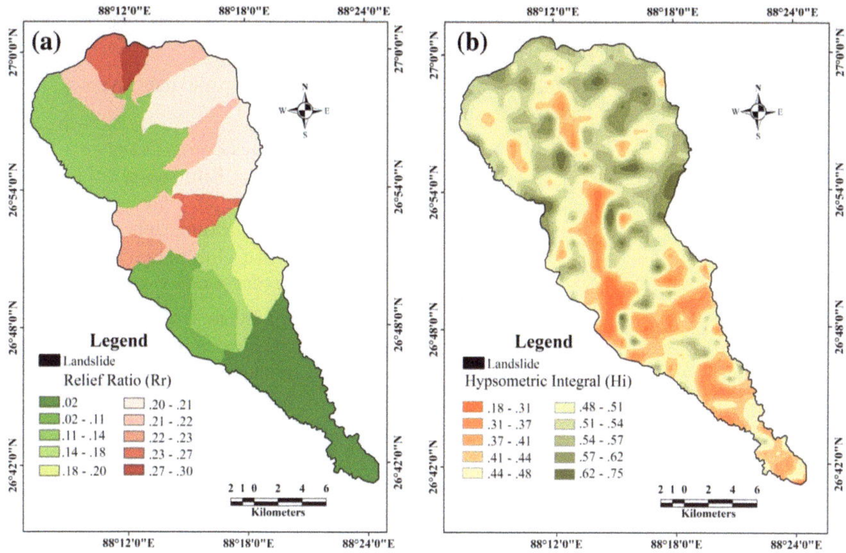

Fig. 4.3 Spatial data layers of **a** relief ratio and **b** hypsometric integral

4.2.2.13 Relief Ratio

Relief ratio is an index of relief characteristics and steepness of a drainage basin. It refers to the difference in height between highest and lowest points in the basin to the horizontal distance along the longest dimension of the basin parallel to the main stream line (Schumm 1956). The ratio is positively correlated with the rate of erosion in the basin and value of relief ratio normally increases with decreasing drainage area and size of a given catchment. In the present study, relief ratio varies from 0.02 to 0.30 and it was classified into ten different classes i.e. 0.02, 0.02–0.11, 0.11–0.14, 0.14–0.18, 0.18–0.20, 0.20–0.21, 0.21–0.22, 0.22–0.23, 0.23–0.27 and 0.27–0.30 (Fig. 4.3a).

4.2.2.14 Hypsometric Integral/Relative Height

Hypsometric related to the measurement and analysis of relationships between altitude and basin area. It helps for understanding the degree of dissection and stage of erosion of a basin (Strahler 1952). The hypsometric integral indicates the percentage area under the dimensionless curve is a good indicative of the basin erosional status or the surface decay whereas the erosional integral is the area proportionate to the area above the curve. The high value of hypsometric integral is denoting the active erosional youthful stage of the basin which is more slides prone (Ghosh 2015; Kumar and Garbyal 2016). The hypsometric integral map of the study area was divided into

Table 4.3 Spatial data layers used in relief diversity, drainage diversity and morphometric diversity for landslide susceptibility mapping

Method	Number of spatial data layers	Name of spatial data layers
Drainage diversity	9	Drainage density, Drainage frequency, Bifurcation ratio, Relative relief, Drainage intensity, Drainage texture, Junction frequency, Infiltration number and Ruggedness index
Relief diversity	8	Slope, Relative relief, Dissection index, Elevation, Ruggedness index, Bifurcation ratio, Relief ratio and Hypsometric integral/Relative height
Morphometric diversity	13	Drainage density, Drainage frequency, Drainage intensity, Dissection index, Drainage texture, Junction frequency, Infiltration number, Bifurcation ratio, Relative relief, Ruggedness index, Elevation, Slope, and Hypsometric integral/Relative height

10 classes i.e. 0.18–0.31, 0.31–0.37, 0.37–0.41, 0.41–0.44, 0.44–0.48, 0.48–0.51, 0.51–0.54, 0.54–0.57, 0.57–0.62 and 0.62–0.75 (Fig. 4.3b).

4.2.3 *Uniform Scaling and Assigning PCA Based Weight for Constructing DD, RD and MD Models and Landslide Susceptibility Zonation Mapping*

After preparing the morphometric data layers in GIS platform, DD, RD and MD models were used for landslide susceptibility zonation mapping (Table 4.3). Some parameters are common because these parameters directly or indirectly represent both the relief and drainage character such as Rb which indicates the drainage characteristics and break of slope characteristics. In the same way Ri implies the arrangement of drainage network and ruggedness characteristics of relief.

To perform the aforesaid three model, principal component analysis (PCA) based weights were assigned to each data layers of each mode. PCA based weight was used by Khatun and Pal (2017). It was calculated using Eq. (4.2).

$$Wm = \frac{\sum amr}{\sum anr\max} \tag{4.2}$$

where Wm is the weight of the *m*th parameter, amr is the row wise summation of correlation coefficient of the mth parameter (avoiding minus symbol), anr max is the maximum correlation coefficient within the total parameter.

The preparation of landslide susceptibility map requires that the factors considered be combined with respect to their relative importance in landslide occurrence (Sarkar and Kanungo 2004; Kanungo et al. 2006). The intra class ranking of 14 parameters was done based on the 10-point scale to make the parameters unidirectional in terms of landslide susceptibility. 10 point was assigned to the class which enhances the chance of landslide occurrences of each data layer (Tables 4.4, 4.5 and 4.6). Except Rb data layer, all the data layers were divided into 10 classes for simplified the work flow. After giving the relative importance on 10 point scale (Tables 4.7, 4.8 and 4.9), each unidirectional and reclassified data layer was multiplied with the individual PCA based parameter weight and last summed up by weighted linear combination method (WLCM).

$$\text{WLCM} = \sum_{j=1}^{n} aj \times wj \tag{4.3}$$

where aj is the unidirectional and reclassified data layer of *j*th attribute, wj is the PCA based weight of the *j*th attribute.

4.2.4 Validation Methods

Two well recognized validation methods were employed in this study i.e. receiver operating characteristics (ROC) curve and frequency ratio (FR). Several researchers used ROC curve for the accuracy assessment of landslide susceptibility map (Bui et al. 2012; Ozdemir and Altural 2013). The ROC curve is the graphical presentation of the sensitivity on the y axis and 1-specificity of the model on the *x* axis (Mohammady et al. 2012). The area under curve (AUC) value of ROC curve varies from 0.5 to 1.0, values below 0.5 indicates a random fit and values close to 1 means excellent prediction accuracy. The range value of AUC was classified into 5 classes which shows prediction accuracy of the map i.e. 0.9–1.0 means excellent, 0.8–0.9 means very good, 0.7–0.8 means good, 0.6–0.7 means average and 0.5–0.6 means poor (Yesilnacar 2005). On the other hand, for the consideration of the ideal LS map, the frequency ratio (FR) value must have a greater value in very high susceptibility zone and it was gradually decreases up to very low susceptibility zone (Pourghasemi et al. 2012; Tsangaratos and Ilia 2016; Mondal and Mandal 2017a, b).

Table 4.4 Logic behind intra sub class ranking of factors on 10 point scale and factors weight in relation to landslide susceptibility of drainage diversity method

Factors	Rank at 10 point scale	Logic applied	Total correlation score (xi)	Factor weighted score (xi/xi max)
Drainage density	10 rank at the highest drainage density class	Higher the drainage density value greater the chance of landslide occurrences	7.27	1.000
Drainage frequency	10 rank at the highest drainage intensity class	Drainage frequency has positive correlation with the landslide potentiality	7.27	1.00
Bifurcation ratio	10 rank at the highest bifurcation ratio class	More landslide susceptibility found in higher bifurcated areas	5.14	0.707
Relative relief	10 rank at the highest relative relief class	Higher relative relief favorable for greater chance of landslide occurrences	5.59	0.769
Drainage intensity	10 rank at the highest drainage intensity class	Drainage intensity directly related to landslide occurrences	4.64	0.638
Drainage texture	10 rank at the highest drainage texture class	More landslides found in higher textural class	7.27	1.000
Junction frequency	10 rank at the highest stream junction frequency class	Higher junction frequency increases greater slope instability	7.06	0.971
Infiltration number	10 rank at the highest Infiltration number class	Higher infiltration number represents more landslide probability	6.48	0.891
Ruggedness index	10 rank at the highest ruggedness index class	Slope instability is more in higher roughness area	7.03	0.967

Table 4.5 Logic behind intra sub class ranking of factors on 10 point scale and factors weight in relation to landslide susceptibility of relief diversity method

Factors	Rank at 10 point scale	Logic applied	Total correlation score (xi)	Factor weighted score (xi/xi max)
Slope	10 rank at the highest slope class	Landslide susceptibility is higher in higher slope areas	4.58	0.811
Relative relief	10 rank at the highest relative relief class	Higher relative relief possesses more chance of landslides	5.65	1.000
Dissection index	10 rank at the highest dissection index class	More landslide phenomena usually occurs in highly dissected areas	3.32	0.588
Elevation	10 rank at the highest elevation class	Elevated areas have greater landslide susceptibility	4.81	0.851
Ruggedness index	10 rank at the highest ruggedness index class	Where ruggedness index is high Slope instability is also high	5.03	0.890
Bifurcation ratio	10 rank at the highest bifurcation ratio class	Higher bifurcated areas have higher landslide susceptibility	4.69	0.831
Relief ratio	10 rank at the highest relief ratio class	More value of relief ratio demarcated more landslide events	4.96	0.878
Hypsometric integral/Relative height	10 rank at maximum integral value	Chance of landslide is higher in more integral value	3.40	0.602

Table 4.6 Logic behind intra sub class ranking of factors on 10 point scale and factors weight in relation to landslide susceptibility of morphometric diversity method

Factors	Rank at 10 point scale	Logic applied	Total correlation score (xi)	Factor weighted score (xi/xi max)
Drainage density	10 rank at the highest drainage density class	Higher the drainage density value greater the chance of landslide occurrences	9.03	1.000
Drainage frequency	10 rank at the highest drainage frequency class	More drainage frequency value enhances the landslide potentiality	8.86	0.981
Drainage intensity	10 rank at the highest drainage intensity class	Greater drainage intensity favorable for more landslide occurrences	6.06	0.671
Dissection index	10 rank at the highest dissection index class	More landslide phenomena usually occurs at highly dissected areas	4.09	0.452
Drainage texture	10 rank at the highest drainage texture class	Frequent landslides found in high textural class	8.86	0.981
Junction frequency	10 rank at the highest stream junction frequency class	Where junction frequency is maximum landslide occurrence is also maximum	8.84	0.935
Infiltration number	10 rank at the highest Infiltration number class	Higher infiltration number represents greater landslide probability	7.55	0.836
Bifurcation ratio	10 rank at the highest bifurcation ratio class	Higher bifurcated ratio leads to the higher landslide susceptibility	6.91	0.765

(continued)

Table 4.6 (continued)

Factors	Rank at 10 point scale	Logic applied	Total correlation score (xi)	Factor weighted score (xi/xi max)
Relative relief	10 rank at the highest relative relief class	Higher relative relief maximizes the chance of landslide	8.08	0.895
Ruggedness index	10 rank at the highest ruggedness index class	Slope instability is high in high roughness area	9.03	1.000
Elevation	10 rank at the highest elevation class	More elevated areas usually poses greater landslide susceptibility	7.65	0.848
Slope	10 rank at the highest slope class	Slope value positively related to landslide susceptibility	6.50	0.720
Hypsometric integral/Relative height	10 rank at the highest integral value	Increasing integral value leads to the more slope instability	4.76	0.527

4.3 Results and Discussion

4.3.1 Relationship Between Morphometric Parameters and Landslide Susceptibility

The relationship between morphometric parameters and landslide susceptibility was made based on estimated FR values. FR values reveal the probability of landslide phenomena in a specific area. As Darjeeling Himalaya exhibits wide range of drainage, relief and morphometric diversity, there is wide ranges of landslide potentiality in diversified mountain topography.

4.3.1.1 Drainage Density, Drainage Frequency, Bifurcation Ratio, Relative Relief and Their Landslide Potentiality

It can be observed that, drainage density in the range of 10.60–15.10 has the highest FR value of 2.94 represents the greater probability of movement of materials down

Table 4.7 Factors weight and sub class ranking of drainage diversity method for landslide susceptibility mapping

Factors	Subclasses	Subclasses rank	Factor weight	Factors	Subclasses	Subclasses rank	Factor weight
Drainage density	0–1.31	1	1.000	Drainage frequency	0–4.91	1	1.000
	1.31–2.61	2			4.91–11.05	2	
	2.61–3.67	3			11.05–15.96	3	
	3.67–4.56	4			15.96–21.27	4	
	4.56–5.39	5			21.27–27.00	5	
	5.39–6.27	6			27.00–33.54	6	
	6.27–7.28	7			33.54–40.90	7	
	7.28–8.58	8			40.90–52.35	8	
	8.58–10.60	9			52.35–70.35	9	
	10.60–15.10	10			70.35–104.30	10	
Bifurcation ratio	1.50–2.56	1	0.707	Relative relief	5.19–51.70	1	0.769
	2.56–3.25	2			51.70–127.00	2	
	3.25–3.71	3			127.00–215.00	3	
	3.71–4.01	5			215.00–292.00	4	
	4.01–4.22	6			292.00–346.00	5	
	4.22–4.52	8			346.00–393.00	6	
	4.52–4.98	9			393.00–434.00	7	
	4.98–5.56	10			434.00–475.00	8	
					475.00–532.00	9	
					532.00–662.00	10	
Drainage intensity	0–1.51	1	0.638	Drainage texture	0–1.23	1	1.000
	1.51–3.02	2			1.23–2.76	2	

(continued)

Table 4.7 (continued)

Factors	Subclasses	Subclasses rank	Factor weight	Factors	Subclasses	Subclasses rank	Factor weight
	3.02–4.16	3			2.76–3.99	3	
	4.06–5.67	4			3.99–5.32	4	
	5.67–8.12	5			5.32–6.75	5	
	8.12–12.10	6			6.75–8.38	6	
	12.10–17.96	7			8.38–10.23	7	
	17.96–25.51	8			10.23–13.10	8	
	25.51–34.58	9			13.10–17.59	9	
	34.58–48.17	10			17.59–26.07	10	
Junction frequency	0–2.38	1	0.971	Infiltration number	0–30.89	1	0.891
	2.38–5.82	2			30.89–80.32	2	
	5.82–9.00	3			80.32–148.28	3	
	9.00–12.45	4			148.28–234.78	4	
	12.45–16.42	5			234.78–339.81	5	
	16.42–20.92	6			339.81–481.92	6	
	20.92–26.74	7			481.92–667.27	7	
	26.74–34.95	8			667.27–883.51	8	
	34.95–46.33	9			883.51–1130.66	9	
	46.33–67.49	10			1130.66–1575.50	10	
Ruggedness index	0–0.34	1	0.967				
	0.34–0.93	2					
	0.93–1.42	3					

(continued)

Table 4.7 (continued)

Factors	Subclasses	Subclasses rank	Factor weight	Factors	Subclasses	Subclasses rank	Factor weight
	1.42–1.83	4					
	1.83–2.22	5					
	2.22–2.63	6					
	2.63–3.12	7					
	3.12–3.74	8					
	3.74–4.48	9					
	4.48–6.35	10					

Table 4.8 Factors weight and sub class ranking of relief diversity method for landslide susceptibility mapping

Factors	Subclasses	Subclasses rank	Factor weight	Factors	Subclasses	Subclasses rank	Factor weight
Slope	0–5.03	1	0.811	Relative relief	5.19–51.70	1	1.000
	5.03–11.30	2			51.70–127.00	2	
	11.30–16.80	3			127.00–215.00	3	
	16.80–21.90	4			215.00–292.00	4	
	21.90–26.60	5			292.00–346.00	5	
	26.60–31.10	6			346.00–393.00	6	
	31.10–35.50	7			393.00–434.00	7	
	35.50–40.60	8			434.00–475.00	8	
	40.60–47.00	9			475.00–532.00	9	
	47.00–71.10	10			532.00–662.00	10	
Dissection index	0.06–0.13	1	0.588	Elevation	106–298	1	0.851
	0.13–0.17	2			298–539	2	
	0.17–0.22	3			539–792	3	
	0.22–0.26	4			792–1030	4	
	0.26–0.31	5			1030–1240	5	
	0.31–0.37	6			1240–1440	6	
	0.37–0.42	7			1440–1650	7	
	0.42–0.47	8			1650–1880	8	
	0.47–0.52	9			1880–2130	9	
	0.52–0.59	10			2130–2600	10	
Ruggedness index	0–0.34	1	0.890	Bifurcation ratio	1.50–2.56	1	0.831

(continued)

Table 4.8 (continued)

Factors	Subclasses	Subclasses rank	Factor weight	Factors	Subclasses	Subclasses rank	Factor weight
	0.34–0.93	2			2.56–3.25	2	
	0.93–1.42	3			3.25–3.71	3	
	1.42–1.83	4			3.71–4.01	5	
	1.83–2.22	5			4.01–4.22	6	
	2.22–2.63	6			4.22–4.52	8	
	2.63–3.12	7			4.52–4.98	9	
	3.12–3.74	8			4.98–5.66	10	
	3.74–4.43	9					
	4.48–6.55	10					
Relief ratio	0.02	1	0.878	Hypsometric integral/relative height	0.18–0.31	1	0.602
	0.02–0.11	2			0.31–0.37	2	
	0.11–0.14	3			0.37–0.41	3	
	0.14–0.18	4			0.41–0.44	4	
	0.18–0.20	5			0.44–0.48	5	
	0.20–0.21	6			0.48–0.51	6	
	0.21–0.22	7			0.51–0.54	7	
	0.22–0.23	8			0.54–0.57	8	
	0.23–0.27	9			0.57–0.62	9	
					0.62–0.75	10	
	0.27–0.30	10					

Table 4.9 Factors weight and sub class ranking of morphometric diversity method for landslide susceptibility mapping

Factors	Subclasses	Subclasses rank	Factor weight	Factors	Subclasses	Subclasses rank	Factor weight
Drainage density	0–1.31	1	1.000	Drainage frequency	0–4.91	1	0.981
	1.31–2.61	2			4.91–11.05	2	
	2.61–3.67	3			11.05–15.96	3	
	3.67–4.56	4			15.96–21.27	4	
	4.56–5.39	5			21.27–27.00	5	
	5.39–6.27	6			27.00–33.54	6	
	6.27–7.28	7			33.54–40.90	7	
	7.28–8.58	8			40.90–52.35	8	
	8.58–10.60	9			52.35–70.35	9	
	10.60–15.10	10			70.35–104.30	10	
Drainage intensity	0–1.51	1	0.671	Dissection index	0.06–0.13	1	0.452
	1.51–3.02	2			0.13–0.17	2	
	3.02–4.16	3			0.17–0.22	3	
	4.06–5.67	4			0.22–0.26	4	
	5.67–8.12	5			0.26–0.31	5	
	8.12–12.10	6			0.31–0.37	6	
	12.10–17.96	7			0.37–0.42	7	
	17.96–25.51	8			0.42–0.47	8	

(continued)

Table 4.9 (continued)

Factors	Subclasses	Subclasses rank	Factor weight	Factors	Subclasses	Subclasses rank	Factor weight
	25.51–34.58	9			0.47–0.52	9	
	34.58–48.17	10			0.52–0.59	10	
Drainage texture	0–1.23	1	0.981	Junction frequency	0–2.38	1	0.935
	1.23–2.76	2			2.38–5.82	2	
	2.76–3.99	3			5.82–9.00	3	
	3.99–5.32	4			9.00–12.45	4	
	5.32–6.75	5			12.45–16.42	5	
	6.75–8.38	6			16.42–20.92	6	
	8.38–10.23	7			20.92–26.74	7	
	10.23–13.10	8			26.74–34.95	8	
	13.10–17.59	9			34.95–46.33	9	
	17.59–26.07	10			46.33–67.49	10	
Infiltration number	0–30.89	1	0.836	Bifurcation ratio	1.50–2.56	1	0.765
	30.89–80.32	2			2.56–3.25	2	
	80.32–148.28	3			3.25–3.71	3	
	148.28–234.78	4			3.71–4.01	5	
	234.78–339.81	5			4.01–4.22	6	
	339.81–481.92	6			4.22–4.52	8	
	481.92–667.27	7			4.52–4.98	9	
	667.27–883.51	8			4.98–5.66	10	

(continued)

Table 4.9 (continued)

Factors	Subclasses	Subclasses rank	Factor weight	Factors	Subclasses	Subclasses rank	Factor weight
	883.51–1130.66	9					
	1130.66–1575.50	10					
Relative relief	5.19–51.70	1	0.895	Ruggedness index	0–0.34	1	1.000
	51.70–127.00	2			0.34–0.93	2	
	127.00–215.00	3			0.93–1.42	3	
	215.00–292.00	4			1.42–1.83	4	
	292.00–346.00	5			1.83–2.22	5	
	346.00–393.00	6			2.22–2.63	6	
	393.00–434.00	7			2.63–3.12	7	
	434.00–475.00	8			3.12–3.74	8	
	475.00–532.00	9			3.74–4.48	9	
	532.00–662.00	10			4.48–6.55	10	
Elevation	106–298	1	0.848	Slope	0–5.03	1	0.720
	298–539	2			5.03–11.30	2	
	539–792	3			11.30–16.80	3	
	792–1030	4			16.80–21.90	4	
	1030–1240	5			21.90–26.60	5	
	1240–1440	6			26.60–31.10	6	
	1440–1650	7			31.10–35.50	7	
	1650–1880	8			35.50–40.60	8	

(continued)

Table 4.9 (continued)

Factors	Subclasses	Subclasses rank	Factor weight	Factors	Subclasses	Subclasses rank	Factor weight
	1880–2130	9			40.60–47.00	9	
	2130–2600	10			47.00–71.10	10	
Hypsometric integral/relative height	0.18–).31	1	0.527				
	0.31–).37	2					
	0.37–).41	3					
	0.41–).44	4					
	0.44–).48	5					
	0.48–).51	6					
	0.51–).54	7					
	0.54–).57	8					
	0.57–).62	9					
	0.62–).75	10					

the slope. Also the drainage density classes of 3.67–4.56 (1.97), 7.28–8.58 (1.85), 6.27–7.28 (1.46), 4.56–5.39 (1.38) and 5.39–6.27 (1.29) have the stronger relationship with landslides events. On the other hand, landslides probability was lesser in the drainage density class of 0–1.31 (0.00) (Table 4.10). In the drainage frequency map, class of 70.35–104.30 having the maximum probability of slope failure events (4.59) followed by 52.35–70.35 (2.76), 11.05–15.96 (2.17), 27.00–33.54 (1.87), 15.96–21.27 (0.99), 21.27–27.00 (0.92), 40.90–52.35 (0.69), 4.91–11.05 (0.36) and 0–4.91 (0.00) respectively (Table 4.10). Bifurcation ratio of 3.71–4.01 has highest FR value of 4.14 representing the maximum chance of landslide occurrences in future containing 42.10% landslide pixels to the total pixels followed by classes of 3.25–3.71 (1.63), 4.01–4.22 (0.82), 4.52–4.98 (0.42) etc. respectively. On the other hand, lowest probability of landslide occurrences found in the class of 1.50–2.56 (0.00) having 0.05% landslide pixels (Table 4.10). In the case of relative relief, class of 532.00–662.00 was registered maximum FR value of 10.33 represents greater probability of landslides phenomena covering 34.51% landslide area followed by 475.00–532.00 (1.69), 434.00–475.00 (1.43) etc. respectively. The lowest landslide probability was found in the classes of 5.19–51.70 (0.00) and 51.70–127.00 (0.00) (Table 4.10). The trend line of FR value reveals that drainage density, drainage frequency and relative relief were positively correlated with slope failure which means that slope instability increasing with increasing the value (Fig. 4.4a, b, d). On the other hand, there was negative relation between bifurcation ratio and landslide occurrences (Fig. 4.4c).

4.3.1.2 Drainage Intensity, Drainage Texture, Junction Frequency, Infiltration Number and Their Role in Slope Instability

Drainage intensity class of 3.02–4.16 has the larger FR value of 1.74 which indicates maximum slope instability condition in future times containing 57.26% landslides. Similarly, the classes of 4.06–5.67 (1.60) and 5.67–8.12 (1.35) have the high probability of sliding down the materials. On the other hand, the classes of 0–1.51 (0.00), 17.96–25.51 (0.00), 25.51–34.58 (0.00) and 34.58–48.17 (0.00) were characterized by minimum probability of slope failure events (Table 4.10). The highest probability was found in the drainage textural class of 17.59–26.07 having FR value of 4.59 followed by 8.38–10.23 (4.37), 13.10–17.59 (2.76), 2.76–3.99 (2.17), 6.75–8.38 (1.87) etc. respectively. The lowest probability of landslide occurrences was found in the class of 0–1.23 (0.00) (Table 4.10). For the junction frequency, the class of 46.33–67.49 has the highest FR value 4.52 indicating strong association with slope failure events. Similarly, the class of 0–2.38 has lowest impact on landslide occurrences with FR value of 0.00 accounting 0.14% of total landslides area (Table 4.10). In the case of infiltration number, the class of 883.51–1130.66 acquired highest degree of FR value of 5.47. Also the classes of 1130.66–1575.50 (3.63), 234.78–339.81 (3.13), 667.27–883.51 (3.11) and 481.92–667.27 have the high probability of landslide occurrences. 0–30.89 class has the least degree (0.01) of associationship with landslides (Table 4.10). In general, the chance of landslide occurrences reduces with

Table 4.10 Relation between landslide factors and landslide occurrences based on frequency ratio (FR) value used in relief, drainage and morphometric diversity method (15 * 15 m pixel)

Factors	Subclasses	Total pixels		Landslide occurrence pixels		FR value (% Plo/%Pt)
		Absolute (Pt)	Percentage (% Pt)	Absolute (Plo)	Percentage (% Plo)	
Drainage density (km/km^2)	0–1.31	338,135	20.10	4	0.07	0.00
	1.31–2.61	197,209	11.72	8	0.14	0.01
	2.61–3.67	239,068	14.21	783	14.27	1.00
	3.67–4.56	284,654	16.92	1830	33.35	1.97
	4.56–5.39	246,882	14.68	1108	20.19	1.38
	5.39–6.27	177,847	10.57	746	13.59	1.29
	6.27–7.28	103,596	6.16	494	9.00	1.46
	7.28–8.58	60,735	3.61	366	6.67	1.85
	8.58–10.60	25,924	1.54	73	1.33	0.87
	10.60–15.10	7951	0.47	76	1.39	2.94
Drainage frequency (no. of stream/km^2)	0–4.91	446,653	26.55	4	0.07	0.00
	4.91–11.05	315,499	18.76	370	6.74	0.36
	11.05–15.96	374,903	22.29	2657	48.42	2.17
	15.96–21.27	260,262	15.47	842	15.35	0.99
	21.27–27.00	131,533	7.82	394	7.19	0.92
	27.00–33.54	73,153	4.35	446	8.13	1.87
	33.54–40.90	42,087	2.50	600	10.94	4.37
	40.90–52.35	27,831	1.65	62	1.14	0.69
	52.35–70.35	6621	0.39	60	1.09	2.76

(continued)

Table 4.10 (continued)

Factors	Subclasses	Total pixels		Landslide occurrence pixels		FR value (% Plo/%Pt)
		Absolute (Pt)	Percentage (% Pt)	Absolute (Plo)	Percentage (% Plo)	
	70.35–104.30	3458	0.21	52	0.94	4.59
Bifurcation ratio	1.50–2.56	258,604	15.37	3	0.05	0.00
	2.56–3.25	196,132	11.66	117	2.14	0.18
	3.25–3.71	335,710	19.96	1784	32.51	1.63
	3.71–4.01	171,049	10.17	2310	42.10	4.14
	4.01–4.22	251,207	14.93	673	12.27	0.82
	4.22–4.52	59,034	3.51	72	1.32	0.38
	4.52–4.98	306,877	18.24	424	7.73	0.42
	4.98–5.66	103,388	6.15	104	1.89	0.31
Relative relief (m/km^2)	5.19–51.70	302,541	17.99	0	0.00	0.00
	51.70–127.00	70,515	4.19	1	0.02	0.00
	127.00–215.00	62,607	3.72	17	0.30	0.08
	215.00–292.00	108,994	6.48	88	1.60	0.25
	292.00–346.00	184,800	10.99	532	9.69	0.88
	346.00–393.00	235,126	13.98	568	10.35	0.74
	393.00–434.00	273,927	16.29	455	8.29	0.51
	434.00–475.00	233,117	13.86	1085	19.78	1.43
	475.00–532.00	154,190	9.17	848	15.46	1.69

(continued)

Table 4.10 (continued)

Factors	Subclasses	Total pixels		Landslide occurrence pixels		FR value (% Plo/%Pt)
		Absolute (Pt)	Percentage (% Pt)	Absolute (Plo)	Percentage (% Plo)	
	532.00–662.00	56,183	3.34	1893	34.51	10.33
Drainage intensity (Din/km^2)	0–1.51	290,451	17.27	3	0.05	0.00
	1.51–3.02	517,865	30.79	782	14.25	0.46
	3.02–4.16	553,415	32.90	3142	57.26	1.74
	4.06–5.67	259,504	15.43	1351	24.62	1.60
	5.67–8.12	41,803	2.49	184	3.36	1.35
	8.12–12.10	11,779	0.70	22	0.41	0.58
	12.10–17.96	3957	0.24	3	0.05	0.23
	17.96–25.51	1541	0.09	0	0.00	0.00
	25.51–34.58	1075	0.06	0	0.00	0.00
	34.58–48.17	610	0.04	0	0.00	0.00
Drainage texture (no. of stream/km)	0–1.23	446,654	26.55	4	0.07	0.00
	1.23–2.76	315,499	18.76	370	6.74	0.36
	2.76–3.99	374,903	22.29	2657	48.42	2.17
	3.99–5.32	260,262	15.47	842	15.35	0.99
	5.32–6.75	131,533	7.82	394	7.19	0.92
	6.75–8.38	73,153	4.35	446	8.13	1.87
	8.38–10.23	42,087	2.50	600	10.94	4.37
	10.23–13.10	27,831	1.65	62	1.14	0.69
	13.10–17.59	6621	0.39	60	1.09	2.76

(continued)

Table 4.10 (continued)

Factors	Subclasses	Total pixels		Landslide occurrence pixels		FR value (% Plo/%Pt)
		Absolute (Pt)	Percentage (% Pt)	Absolute (Plo)	Percentage (% Plo)	
	17.59–26.07	3457	0.21	52	0.94	4.59
Junction frequency (no. of stream node/km^2)	0–2.38	500,275	29.74	8	0.14	0.00
	2.38–5.82	325,499	19.35	223	4.06	0.21
	5.82–9.00	376,690	22.40	3229	58.84	2.63
	9.00–12.45	237,366	14.11	616	11.22	0.80
	12.45–16.42	117,809	7.00	312	5.69	0.81
	16.42–20.92	64,700	3.85	713	13.00	3.38
	20.92–26.74	40,209	2.39	241	4.39	1.84
	26.74–34.95	11,322	0.67	44	0.80	1.19
	34.95–46.33	5084	0.30	57	1.03	3.41
	46.33–67.49	3046	0.18	45	0.82	4.52
Infiltration number (In/km^2)	0–30.89	604,163	35.92	14	0.25	0.01
	30.89–80.32	522,448	31.06	2998	54.64	1.76
	80.32–148.28	334,796	19.90	1001	18.25	0.92
	148.28–234.78	121,494	7.22	609	11.10	1.54
	234.78–339.81	54,888	3.26	561	10.23	3.13
	339.81–481.92	30,754	1.83	164	2.99	1.63
	481.92–667.27	5730	0.34	39	0.71	2.09
	667.27–883.51	3757	0.22	38	0.69	3.11
	883.51–1130.66	2572	0.15	46	0.84	5.47

(continued)

Table 4.10 (continued)

Factors	Subclasses	Total pixels		Landslide occurrence pixels		FR value (% Plo/%Pt)
		Absolute (Pt)	Percentage (% Pt)	Absolute (Plo)	Percentage (% Plo)	
	1130.66–1575.50	1399	0.08	17	0.30	3.63
Ruggedness index (Ri/km^2)	0–0.34	398,329	23.68	4	0.07	0.00
	0.34–0.93	139,876	8.32	19	0.34	0.04
	0.93–1.42	234,658	13.95	158	2.88	0.21
	1.42–1.83	252,886	15.03	467	8.50	0.57
	1.83–2.22	265,139	15.76	1537	28.01	1.78
	2.22–2.63	181,650	10.80	2093	38.14	3.53
	2.63–3.12	118,089	7.02	600	10.94	1.56
	3.12–3.74	50,005	2.97	466	8.48	2.85
	3.74–4.48	30,193	1.80	136	2.47	1.38
	4.48–6.55	11,176	0.66	9	0.16	0.24
Dissection index (Di/km^2)	0.06–0.13	222,223	13.21	71	1.30	0.10
	0.13–0.17	153,770	9.14	716	13.06	1.43
	0.17–0.22	206,217	12.26	146	2.67	0.22
	0.22–0.26	282,596	16.80	264	4.82	0.29
	0.26–0.31	272,965	16.23	403	7.35	0.45
	0.31–0.37	155,687	9.26	205	3.74	0.40
	0.37–0.42	133,237	7.92	573	10.44	1.32
	0.42–0.47	125,220	7.44	1192	21.72	2.92
	0.47–0.52	107,111	6.37	1849	33.71	5.29

(continued)

Table 4.10 (continued)

Factors	Subclasses	Total pixels		Landslide occurrence pixels		FR value (% Plo/%Pt)
		Absolute (Pt)	Percentage (% Pt)	Absolute (Plo)	Percentage (% Plo)	
	0.52–0.59	22,974	1.37	66	1.21	0.89
Elevation (m)	106–298	368,681	21.92	4	0.08	0.00
	298–539	139,536	8.30	270	4.93	0.59
	539–792	135,130	8.03	876	15.96	1.99
	792–1030	172,562	10.26	1240	22.60	2.20
	1030–1240	181,438	10.79	1457	26.55	2.46
	1240–1440	189,984	11.30	573	10.44	0.92
	1440–1650	169,253	10.06	273	4.97	0.49
	1650–1880	144,806	8.61	75	1.37	0.16
	1880–2130	118,740	7.06	180	3.29	0.47
	2130–2600	61,869	3.68	539	9.82	2.67
Slope (°)	0-5.03	400,086	23.79	47	0.86	0.04
	5.03–11.30	167,825	9.98	202	3.68	0.37
	11.30–16.80	219,531	13.05	522	9.52	0.73
	16.80–21.90	231,356	13.75	736	13.41	0.98
	21.90–26.60	207,903	12.36	601	10.95	0.89
	26.60–31.10	169,058	10.05	625	11.38	1.13
	31.10–35.50	126,742	7.54	611	11.14	1.48
	35.50–40.60	91,415	5.43	663	12.09	2.22
	40.60–47.00	51,416	3.06	742	13.52	4.42

(continued)

Table 4.10 (continued)

Factors	Subclasses	Total pixels		Landslide occurrence pixels		FR value (% Plo/%Pt)
		Absolute (Pt)	Percentage (% Pt)	Absolute (Plo)	Percentage (% Plo)	
	47.00–71.10	16,667	0.99	738	13.44	13.57
Relief ratio	0.02	258,604	15.37	3	0.05	0.00
	0.02–0.11	122,491	7.28	94	1.71	0.23
	0.11–0.14	435,415	25.89	217	3.95	0.15
	0.14–0.18	82,960	4.93	642	11.70	2.37
	0.18–0.20	92,171	5.48	23	0.43	0.08
	0.20–0.21	261,419	15.54	1718	31.32	2.01
	0.21–0.22	272,920	16.23	2747	50.06	3.09
	0.22–0.23	34,196	2.03	1	0.02	0.01
	0.23–0.27	96,540	5.74	31	0.57	0.10
	0.27–0.30	25,284	1.50	11	0.20	0.13
Hypsometric integral/Relative height	0.18–0.31	24,904	1.48	28	0.52	0.35
	0.31–0.37	73,460	4.37	95	1.72	0.39
	0.37–0.41	133,549	7.94	178	3.23	0.41
	0.41–0.44	196,047	11.66	350	6.38	0.55
	0.44–0.48	269,836	16.04	762	13.88	0.87
	0.48–0.51	305,224	18.15	542	9.88	0.54
	0.51–0.54	302,130	17.96	961	17.51	0.97
	0.54–0.57	234,590	13.95	1127	20.55	1.47
	0.57–0.62	111,499	6.63	1085	19.77	2.98
	0.62–0.75	30,762	1.83	360	6.56	3.59

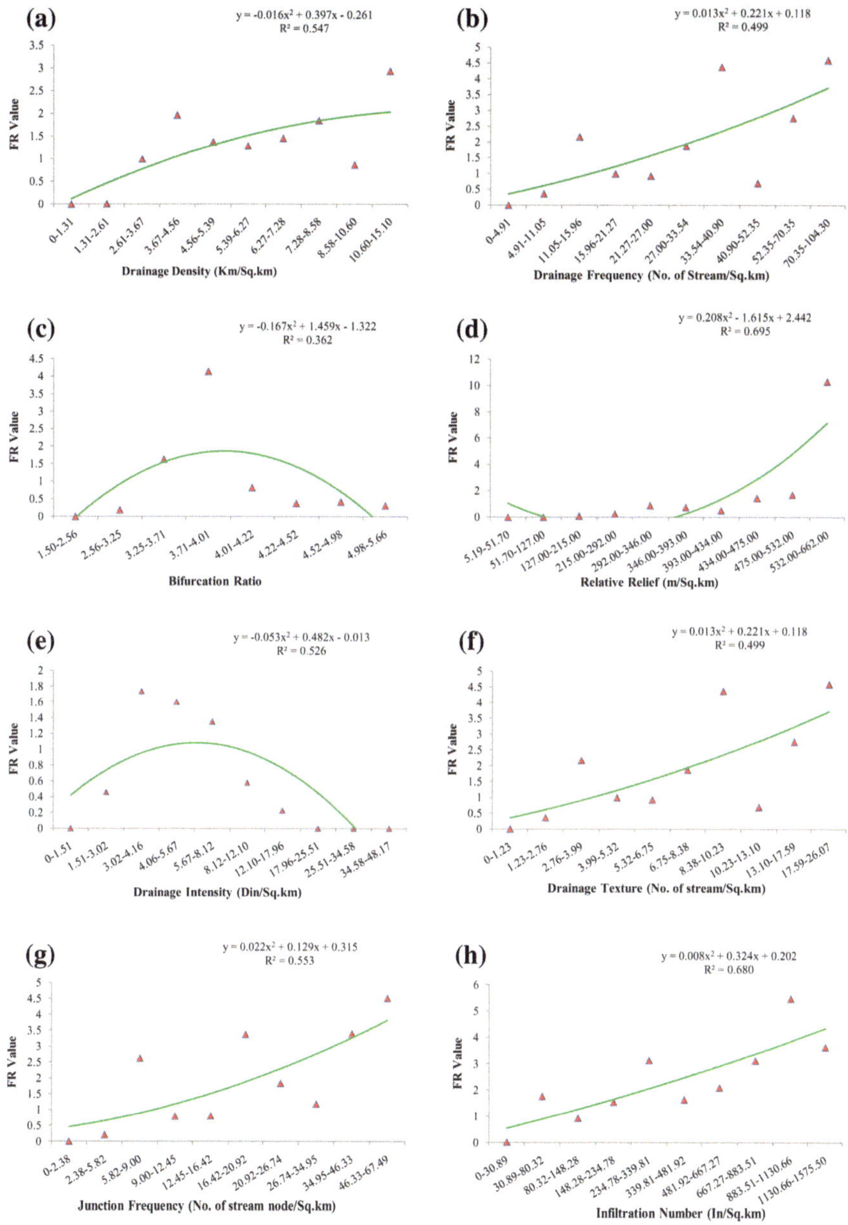

Fig. 4.4 Relation between data layers and FR value **a** drainage density, **b** drainage frequency, **c** bifurcation ratio, **d** relative relief, **e** drainage intensity, **f** drainage texture, **g** junction frequency and **h** infiltration number

increasing drainage intensity (Fig. 4.4e). On the other hand, drainage texture, Junction frequency and infiltration number were positively linked with slope failure (Fig. 4.4f, g, h).

4.3.1.3 Ruggedness Index, Dissection Index, Altitude, Slope and Their Role in Slope Failure

For ruggedness index, the highest FR value of 3.53 was obtained in the class of 2.22–2.63 which has the greater potentiality of slope failures. The classes of 3.12–3.74 (2.85), 1.83–2.22 (1.78), 2.63–3.12 (1.56) and 3.74–4.48 (1.38) also have the stronger relationship with landslides. The least probability of landslide occurrences was found in the class of 0–0.34 (0.00) (Table 4.10). In the case of dissection index, the class of 0.47–0.52 acquired highest degree of FR value of 5.29 covering 33.71% landslides area to total landslides area followed by 0.42–0.47 (2.92), 0.13–0.17 (1.43), 0.37–0.42 (1.32), 0.52–0.59 (0.89), 0.26–0.31 (0.45), 0.31–0.37 (0.40), 0.22–0.26 (0.29), 0.17–0.22 (0.22) and 0.06–0.13 (0.10) respectively (Table 4.10). It can be observed that, elevation in the range of 2130–2600 m has the highest FR value of 2.67 represents the greater probability of movement of materials down the slope. Also the classes of 1030–1240 m (2.46), 792–1030 m (2.20) and 539–792 m (1.99) have the stronger relationship with landslides events. On the other hand, landslides probability was lesser in the elevation class of 106–298 m (0.00) (Table 4.10). Sl angle of 47.00°–71.10° has the highest FR value of 13.57 represents maximum chance of landslide occurrences in future containing 13.44% landslide pixels to the total pixels followed by slope ranges of 40.60°–47.00° (4.42), 35.50°–40.60° (2.22), 22.03°–25.03° (1.48) etc. respectively. On the other hand, lowest probability of landslide occurrences was found in the class of 0°–5.03° (0.04) slope range (Table 4.10). In general, all these 4 factors were positively associated with slope instability (Fig. 4.5a–d). If the value of the factors increases, the chance of landslide also increases.

4.3.1.4 Relief Ratio, Hypsometric Integral and Their Role in Landslide Potentiality

In the relief ratio layer, the highest value of FR (3.09) was observed in the class of 0.21–0.22 indicating greater probability of landslide occurrences. The other classes of 0.14–0.18 (2.37) and 0.20–0.21 (2.01) also have high probability of slope failure. On the other hand, lowest probability of landslides was observed in the class of 0.02 with FR value of 0.00 (Table 4.10). The FR value of hypsometric integral showed that, the class of 0.62–0.75 has the most effect on the incidence of landslides with FR value of 3.59 followed by the classes of 0.57–0.62 (2.98), 0.54–0.57 (1.47), 0.51–0.54 (0.97), 0.44–0.48 (0.87), 0.41–0.44 (0.55), 0.48–0.51 (0.54), 0.37–0.41 (0.41), 0.31–0.37 (0.39) and 0.18–0.31 (0.35) respectively (Table 4.10). Generally,

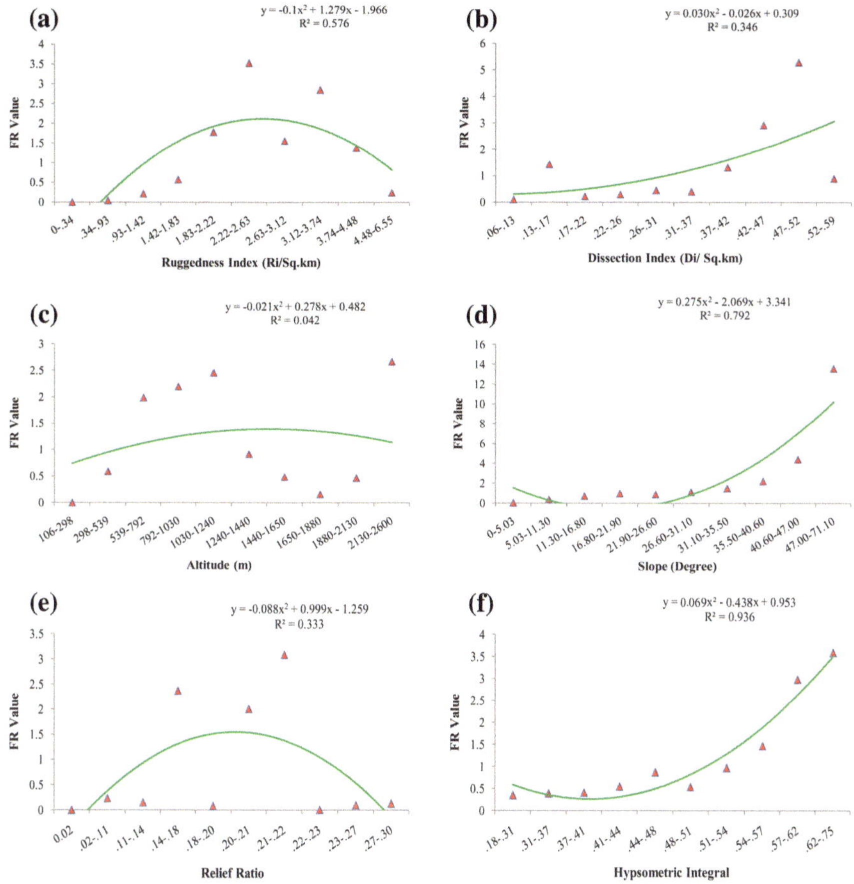

Fig. 4.5 Relation between data layers and FR value **a** ruggedness index, **b** dissection index, **c** altitude, **d** slope, **e** relief ratio and **f** hypsometric integral

the chance of landslide occurrences increases with the increasing value of relief ratio and hypsometric integral (Fig. 4.5e, f).

4.3.2 Landslide Susceptibility Mapping Using DD, RD and MD Models

Landslide susceptibility maps based on the three different models were prepared using PCA based weight and their composite equations were:

$$
\begin{aligned}
\text{DD Model} = {} & (\text{“Reclass of Drainage Density”} \times 1.000) \\
& + (\text{“Reclass of Drainage Frequency”} \times 1.000) \\
& + (\text{“Reclass of Bifurcation Ratio”} \times 0.707) \\
& + (\text{“Reclass of Relative relief”} \times 0.769) \\
& + (\text{“Reclass of Drainage Intensity”} \times 0.638) \\
& + (\text{“Reclass of Drainage Texture”} \times 1.000) \\
& + (\text{“Reclass of Junction Frequency”} \times 0.971) \\
& + (\text{“Reclass of Infiltration Number”} \times 0.891) \\
& + (\text{“Reclass of Ruggedness Index”} \times 0.967)
\end{aligned} \tag{4.4}
$$

$$
\begin{aligned}
\text{RD Model} = {} & (\text{“Reclass of Slope”} \times 0.811) \\
& + (\text{“Reclass of Relative relief”} \times 1.000) \\
& + (\text{“Reclass of Dissection Index”} \times 0.588) \\
& + (\text{“Reclass of Altitude”} \times 0.851) \\
& + (\text{“Reclass of Ruggedness Index”} \times 0.890) \\
& + (\text{“Reclass of Bifurcation Ratio”} \times 0.831) \\
& + (\text{“Reclass of Relief ratio”} \times 0.878) \\
& + (\text{“Reclass of Hypsometric Integral”} \times 0.602)
\end{aligned} \tag{4.5}
$$

$$
\begin{aligned}
\text{MD Model} = {} & (\text{“Reclass of Drainage Density”} \times 1.000) \\
& + (\text{“Reclass of Drainage Frequency”} \times 0.981) \\
& + (\text{“Reclass of Drainage Intensity”} \times 0.671) \\
& + (\text{“Reclass of Dissection Index”} \times 0.452) \\
& + (\text{“Reclass of Drainage Texture”} \times 0.981) \\
& + (\text{“Reclass of Junction Frequency”} \times 0.935) \\
& + (\text{“Reclass of Infiltration Number”} \times 0.836) \\
& + (\text{“Reclass of Bifurcation Ratio”} \times 0.765) \\
& + (\text{“Reclass of Relative relief”} \times 0.895) \\
& + (\text{“Reclass of Ruggedness Index”} \times 1.000) \\
& + (\text{“Reclass of Altitude”} \times 0.848) \\
& + (\text{“Reclass of Slope”} \times 0.720) \\
& + (\text{“Reclass of Hypsometric Integral”} \times 0.527)
\end{aligned} \tag{4.6}
$$

DD model was prepared for identifying highly diversified drainage area where geomorphic process is actively engaged in landform modification whereas RD model was constructed to find out the probable areas of landslide based on the relief configuration parameters. Finally, MD model was prepared by using drainage and relief parameters. In this study, the values of landslide susceptibility map using DD model was varies from 7.94 to 88.46. Higher value indicates greater possibility of landslide occurrences and vice versa. For visual interpretation, continuous landslide susceptibility value was classified into five different categorical landslides susceptibility zones using natural breaks classification method (Bednarik et al. 2010; Constantin et al. 2011) i.e. very low (7.94–16.47), low (16.47–27.52), moderate (27.52–36.68), high (36.68–47.73) and very high (47.73–88.46) (Fig. 4.6a) covering an area of 22.67, 15.88, 31.16, 22.07 and 8.22% respectively of the total area of Balason river basin (Table 4.11). It was seen that, the 77.14% landslide pixels was found in very high and high susceptibility zones while the minimum figure was lies in the class of very low susceptibility zones (0.05%) (Table 4.11). Similarly, both the landslide susceptibility values of RD and MD models were divided into five different classes using natural breaks classification method (Figs. 4.6b, c). The very low, low, moderate, high and very high susceptibility zones covering an area of 17.50, 9.61, 22.10, 33.30 and 17.49% respectively using RD model. In case of MD model, these zones covered 22.57, 15.11, 31.98, 22.36 and 7.98% respectively (Table 4.11). The proportion of landslide pixels to total landslide pixels was decreasing from very high to very low susceptibility zones using RD model. About 70% landslide pixels were confined in the very low to moderate susceptibility classes using MD model (Table 4.11).

4.3.3 Validation and Accuracy Assessment

The area under curve (AUC) value of receiver operating characteristics (ROC) curve indicated that the prediction accuracy of the DD, RD and MD models was 71.4, 73.9 and 76.3% respectively (Fig. 4.7a). The FR plot of the aforementioned three models indicated that the chance of landslide occurrences is gradually increasing from very low to very high susceptibility class (Fig. 4.7b and see Table 4.11). The two-validation procedures showed that the DD, RR and MD models yield reasonable prediction accuracy and can be used for identifying the role of drainage, relief and morphometric parameters in landslide susceptibility assessment. It can be mentioned here, the findings of Mondal and Mandal (2017a, b) work on Balason river basin suggested that, the prediction accuracy of the landslide susceptibility maps was higher using frequency ratio (FR) and logistic regression (LR) models (94.20 and 96.10% respectively). They incorporated several geomorphological, lithological, hydrological, protective, anthropogenic and triggering factors in the preparation of landslide susceptibility maps.

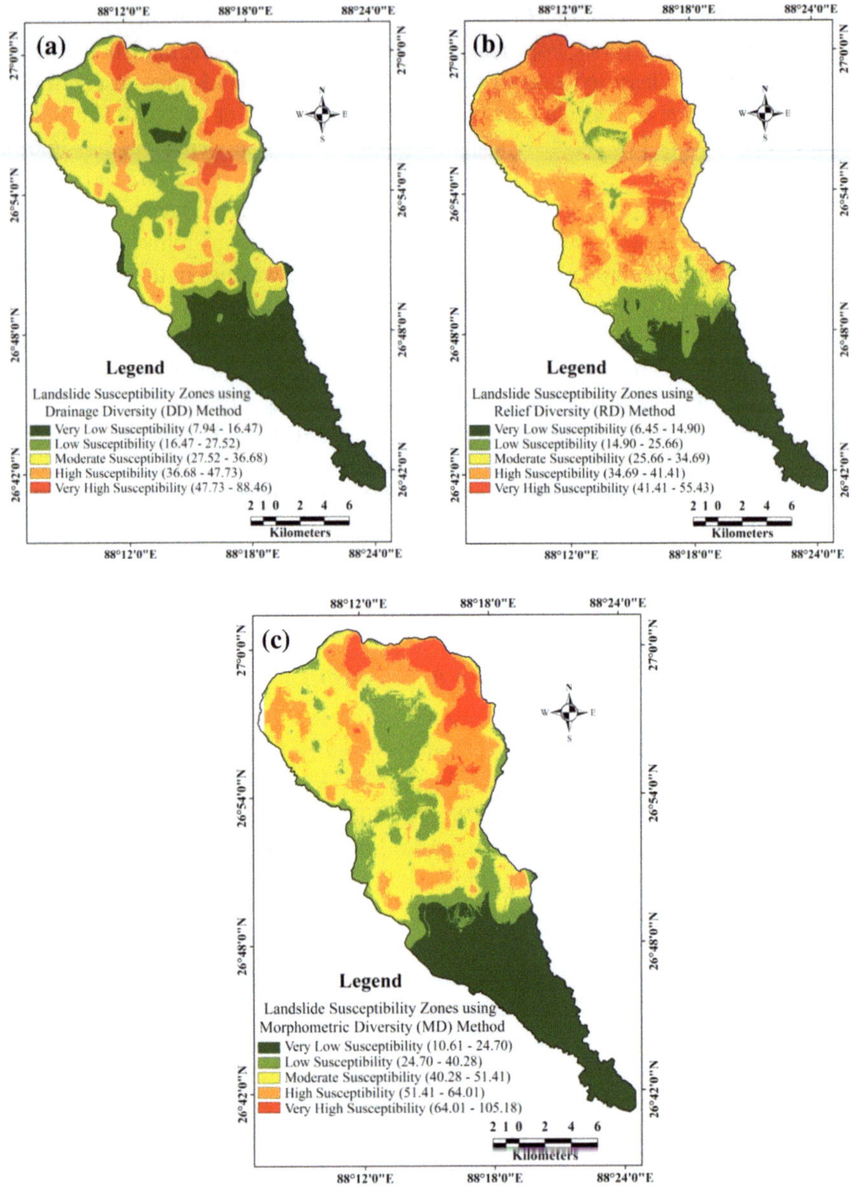

Fig. 4.6 Landslide susceptibility zonation mapping using **a** DD, **b** RD and **c** MD methods

Table 4.11 Statistics of different susceptibility classes of the prepared landslide susceptibility maps using relief diversity (RD), drainage diversity (DD) and morphometric diversity (MD) methods

Landslide susceptibility zones			Very low	Low	Moderate	High	Very high
Using DD method	Total pixels	Absolute	381,311	267,158	524,092	371,199	138,240
		Percentage	22.67	15.88	31.16	22.07	8.22
	Landslide occurrence pixels	Absolute	3	195	1056	3000	1233
		Percentage	0.05	3.56	19.25	54.67	22.47
	Frequency ratio (FR) value		0.00	0.22	0.62	2.48	2.73
Using RD method	Total pixels	Absolute	294,320	161,612	371,767	560,142	294,159
		Percentage	17.50	9.61	22.10	33.30	17.49
	Landslide occurrence pixels	Absolute	1	49	219	2087	3132
		Percentage	0.02	0.89	3.98	38.03	57.08
	Frequency ratio (FR) value		0.00	0.09	0.18	1.14	3.26
Using MD method	Total pixels	Absolute	379,664	254,090	537,934	376,104	134,208
		Percentage	22.57	15.11	31.98	22.36	7.98
	Landslide occurrence pixels	Absolute	3	124	1121	3000	1233
		Percentage	0.05	2.26	20.43	54.67	22.47
	Frequency ratio (FR) value		0.00	0.15	0.64	2.45	2.82

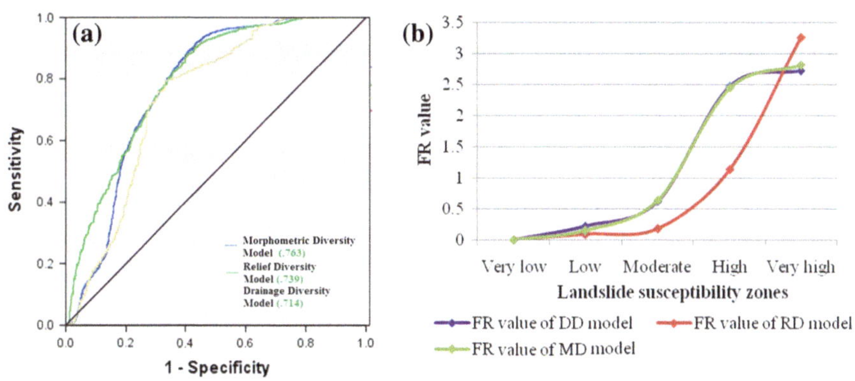

Fig. 4.7 Validation of the Landslide susceptibility maps using landslide locations **a** ROC curve and **b** frequency ratio plot

4.3.4 Limitations of the Prepared Landslide Susceptibility Maps

The prepared landslide susceptibility maps do not define the chronological characteristics of the slope failure events and detailed risk zones. The susceptibility maps should be viewed at the initial landslide mapping. This study does not delineate the areas which become unstable during earthquake. In addition, the limitations of the study also include incomplete landslide inventory map due to the restriction of air photos and lack of information of past landslides, generation of errors to the model due to the scale and mapping roughness of the landslide causative factors.

4.4 Conclusion

It can be concluded that each model can be treated as unique in terms of terrain processes and forms and can be used to identify role of drainage and relief in slope failures individually. Only drainage diversity model predicts 71.40% landslides whereas relief diversity predicts 73.9% landslides. The landslide prediction is slightly better when drainage and relief diversity are combine together. Morphometric diversity (combination of relief and drainage diversity) has the area ratio 0.763 and predicts 76.3% landslides. However, one must be careful while using these geomorphic diversity models for specific site development because of the scale of analysis where other slope factors need to be considered.

References

Akgun A, Sezer EA, Nefesliogl HA, Gokceoglu C, Pradhan B (2012) An easy-to-use MATLAB program (MamLand) for the assessment of landslide susceptibility using a Mamdani fuzzy algorithm. Comput Geosci 38:23–34

Althuwaynee OF, Pradhan B, Park H, Lee JH (2014) A novel ensemble bivariate statistical evidential belief function with knowledge-based analytical hierarchy process and multivariate statistical logistic regression for landslide susceptibility mapping. Catena 114:21–36

Amani M, Safaviyan A (2015) Sub-basins prioritization using morphometric analysis- remote sensing technique and GIS-Golestan-Iran. Int Lett Nat Sci 38:56–65

Anbalagan R (1992) Landslide hazard evaluation and zonation mapping in mountainous terrain. Eng Geol 32:269–277

Ardizzone F, Cardinali M, Carrara A, Guzzetti F, Reichenbach P (2002) Impact of mapping errors on the reliability of landslide hazard maps. Nat Hazards Earth Syst Sci 2:3–14

Avinash K, Deepika B, Jayappa KS (2014) Basin geomorphology and drainage morphometry parameters used as indicators for groundwater prospect: insight from geographical information system (GIS) technique. J Earth Sci 25(6):1018–1032. https://doi.org/10.1007/s12583-014-0505-8

Ayalew L, Yamagishi H, Marui H, Kanno T (2005) Landslides in Sado island of Japan: part II, GIS-based susceptibility mapping with comparisons of results from two methods and verifications. Eng Geol 81:432–445
Bui DT, Pradhan B, Lofman O, Revhaug I, Dick OB (2012) Spatial prediction of landslide hazards in Hoa Binh province (Vietnam): a comparative assessment of the efficacy of evidential belief functions and fuzzy logic models. CATENA 96:28–40
Chau KT, Sze YL, Fung MK, Wong WY, Fong EL, Chan LCP (2004) Landslide hazard analysis for Hong Kong using landslide inventory and GIS. Comput Geosci 30:429–443
Choi J, Oh HJ, Lee HJ, Lee C, Lee S (2011) Combining landslide susceptibility maps obtained from frequency ratio logistic regression and artificial neural network models using aster images and GIS. Eng Geol 124:12–23
Chung C-JF, Fabbri AG (1999) Probabilistic prediction models for landslide hazard mapping. Photogram Eng Remote Sens 65(12):1389–1399
Dai FC, Lee CF (2002) Landslide characteristics and slope instability modeling using GIS Lantau Island Hong Kong. Geomorphology 42:213–228
Devkota KC, Regmi AD, Pourghasemi HR et al (2013a) Landslide susceptibility mapping using certainty factor, index of entropy and logistic regression models in GIS and their comparison at Mugling-Narayanghat road section in Nepal Himalaya. Nat Hazards 65(1):135–165. https://doi.org/10.1007/s11069-012-0347-6
Devkota KC, Regmi AD, Pourghasemi HR, Yoshida K, Pradhan B, Ryu IC, Dhital MR, Althuwaynee OF (2013b) Landslide susceptibility mapping using certainty factor index of entropy and logistic regression models in GIS and their comparison at Mugling-Narayanghat road section in Nepal Himalaya. Nat Hazards 65:135–165
Dove N (1957) The ratio of relative and absolute altitude of Mt.Camel. Geog Rev 47:564–569
Faniran A (1968) The index of drainage intensity—a provisional new drainage factor. Aust J Sci 31:328–330
Farhan Y, Anbar A, Enaba O, Al-Shaikh N (2015) Quantitative analysis of geomorphometric parameters of Wadi Kerak, Jordan, using remote sensing and GIS. J Water Resour Protect 7:456–475. https://doi.org/10.4236/jwarp.2015.76037
Foumelis M, Lekkas E, Parcharidis I (2004) Landslide susceptibility mapping by GIS-based qualitative weighting procedure in Corinth area. Bull Geol Soc Greece XXXVI
Gajbhiye S, Mishra SK, Pandey A (2014) Prioritizing erosion-prone area through morphometric analysis: an RS and GIS perspective. Appl Water Sci 4(1):51–61
Ghosh D (2015) Landslide susceptibility analysis from morphometric parameter analysis of Riyong Khola basin, West Sikkim, India: a geospatial approach. Int J Geol 5(1):54–65
Ghosh KG, Saha S (2015) Identification of soil erosion susceptible areas in Hinglo river basin, Eastern India based on geo-statistics. Univers J Environ Res Technol 5(3)
Ghosh S, Carranza EJM, van Westen CJ, Jetten VG, Bhattacharya DN (2011) Selecting and weighting spatial predictors for empirical modeling of landslide susceptibility in the Darjeeling Himalayas (India). Geomorphology 131(1–2):35–56. https://doi.org/10.1016/j.geomorph.2011.04.019
Gopal KG, Saha S (2015) Identification of soil erosion susceptible areas in Hinglo river basin, Eastern India based on geostatistics. Univ J Environ Res Technol 5(3):152–164
Gupta RP, Joshi BC (1990) Landslide Hazard Zonation using the GIS Approach—a case Study from the Ramganga Catchment Himalayas. Eng Geol 28:119–131
Horton RE (1932) Drainage basin characteristics. Am Geophys Union 13:350–361
Horton RE (1945) Erosional development of streams and their drainage basins, a hydrophysical approach to quantitative morphology. Geol Soc Am Bull 56:275–370
Jeganathan C, Chauniyal DD (2000) An evidential weighted approach for landslide hazard zonation from geo-environmental characterization: a case study of Kelani area. Curr Sci 79(2):238–243

Kanungo DP, Arora MK, Sarkar S, Gupta RP (2006) A comparative study of conventional, ANN black box, fuzzy and combined neural and fuzzy weighting procedures for landslide susceptibility Zonation in Darjeeling Himalayas. Eng Geol 85:347–366
Karim S, Jalileddin S, Ali MT (2011) Zoning landslide by use of frequency ratio method (case study: Deylaman Region). Middle East J Sci Res 9(5):578–583
Khatun S, Pal S (2017) Categorization of morphometric surface through morphometric diversity analysis in Kushkarani River basin of Eastern India. Asian J Phys Chem Sci 2(1):1–19. https://doi.org/10.9734/ajopacs/2017/31098
Kienholz H (1978) Maps of geomorphology and natural hazards of Grindelwald, Switzerland: scale 1:10,000. Arct Alp Res 10:169–184
Kumar K, Garbyal Y (2016) Analysis of morphometric parameters for the identification of probable landslide occurrences. In: Conference: Geo-Chicago, 14–16 Aug 2016, ASCE library, subject heading- Himalayas, Chicago, Illinois At. https://doi.org/10.1061/9780784480120.035
Lee S, Pradhan B (2007) Landslide hazard mapping at Selangor Malaysia using frequency ratio and logistic regression models. Landslides 4:33–41
Lee S, Sambath T (2006) Landslide susceptibility mapping in the Damrei Romel area Cambodia using frequency ratio and logistic regression models. Environ Geol 50:847–855
Lee S, Talib JA (2005) Probabilistic landslide susceptibility and factor effect analysis. Environ Geol 47:982–990
Magesh NS, Jitheshlal K, Chandrasekar N, Jini K (2012) GIS based morphometric evaluation of Chimini and Mupily watersheds, parts of Western Ghats, Thrissur district, Kerala, India. Earth Sci Inform 5:111–121
Mahalingam R, Olsen MJ, O'Banion MS (2016) Evaluation of landslide susceptibility mapping techniques using lidar–derived conditioning factors (Oregon case study). Geomat Nat Hazard Risk 7:1884–1907
Majtan S, Omura H, Morita K (2002) Fractal dimension as an indicator of probability for landslides in North Matsuura Japan. Geograficky Casopis 54:5–19
Mandal B, Mandal S (2016) Assessment of mountain slope instability in the Lish river basin of Eastern Darjeeling Himalaya using frequency ratio model (FRM). Earth Syst Environ 2:121. https://doi.org/10.1007/s40808-016-0169-8
Mandal S, Mandal K (2017) Bivariate statistical index for landslide susceptibility mapping in the Rorachu River basin of Eastern Sikkim Himalaya, India. Spat Inf Res. https://doi.org/10.1007/s41324-017-0156-9
Mandal S, Mandal K (2018) Modeling and mapping landslide susceptibility zones using GIS based multivariate binary logistic regression (LR) model in the Rorachu river basin of eastern Sikkim Himalaya, India. Model Earth Syst Environ. https://doi.org/10.1007/s40808-018-0426-0
Miller CT, Poirier-McNeill MM, Mayer AS (1990) Dissolution of trapped nonaqueous phase liquids: mass transfer characteristics. Water Resour Res 26:2783–2796
Mohammady M, Pourghasemi HR, Pradhan B (2012) Landslide susceptibility mapping at Golestan Province, Iran: a comparison between frequency ratio, Dempster-Shafer, and weights-of-evidence models. J Asian Earth Sci 61:221–236
Mondal S, Mandal S (2017a) RS & GIS-based landslide susceptibility mapping of the Balason River basin, Darjeeling Himalaya, using logistic regression (LR) model. Georisk Assess Manag Risk Eng Syst Geohazards. https://doi.org/10.1080/17499518.2017.1347949
Mondal S, Mandal S (2017b) Application of frequency ratio (FR) model in spatial prediction of landslides in the Balason river basin, Darjeeling Himalaya. Spat Inf Res. https://doi.org/10.1007/s41324-017-0101-y
Nag SK (1998) Morphometric analysis using remote sensing techniques in the Chaka sub-basin, Purulia district, West Bengal. J Indian Soc Remote Sens 26(1):69–76
Nautiyal MD (1994) Morphometric analysis of drainage basin, district Dehradun, Uttar Pradesh. J Indian Soc Remote Sens 22(1994):252–262

Nefeslioglu HA, Sezer E, Go¨kc¸eog˘lu C, Bozkır AS, Duman TY (2010) Assessment of landslide susceptibility by decision trees in the metropolitan area of Istanbul Turkey. Math Prob Eng (Article ID: 901095)

Nidhi K, Chowdary VM, Tiwari KN, Shinde V, Dadhwal VK (2016) Assessment of surface water potential using morphometry and curve number-based approaches. Geocarto Int 32(11):1206–1228. https://doi.org/10.1080/10106049.2016.1195889

Ozdemir A, Altural T (2013) A comparative study of frequency ratio, weights of evidence and logistic regression methods for landslide susceptibility mapping: Sultan Mountains, SW Turkey. J Asian Earth Sci 64:180–197

Poli S, Sterlacchini S (2007) Landslide representation strategies in susceptibility studies using weights-of-evidence modeling technique. Nat Res Res 16(2). https://doi.org/10.1007/s11053-007-9043-8

Pourghasemi HR, Pradhan B, Gokceoglu C, Mohammadi M, Moradi HR (2012) Application of weights-of-evidence and certainty factor models and their comparison in landslide susceptibility mapping at Haraz watershed, Iran. Arab J Geosci. https://doi.org/10.1007/s12517-012-0532-7

Pradhan B (2010) Use of GIS-based fuzzy logic relations and its cross application to produce landslide susceptibility maps in three test areas in Malaysia. Environ Earth Sci 63:329–349

Pradhan B, Lee S (2010a) Delineation of landslide hazard areas on Penang Island, Malaysia, by using frequency ratio, logistic regression, and artificial neural network models. Environ Earth Sci 60(5):1037–1054

Pradhan B, Lee S (2010b) Landslide susceptibility assessment and factor effect analysis: backpropagation artificial neural networks and their comparison with frequency ratio and bivariate logistic regression modelling. Environ Model Softw 25(6):747–759

Rai PK, Mishra S, Ahmad A, Mohan K (2014) A GIS-based approach in drainage morphometric analysis of Kanhar River Basin, India. Appl Water Sci. https://doi.org/10.1007/s13201-014-0238-y

Rastogi RA, Sharma TC (1976) Quantitative analysis of drainage basin characteristics. J Soil Water Conserv India 26(1&4):18–25

Sarkar S, Kanungo DP (2004) An integrated approach for landslide susceptibility mapping using remote sensing and GIS. Photogram Eng Remote Sens 70(5):617–625

Schumm SA (1956) Evolution of drainage systems and slopes in badlands at Perth Amboy, New Jersey. Bull Geol Soc Am 67:597

Shrestha S, Kang T, Suwal MS (2017) An ensemble model for co-seismic landslide susceptibility using GIS and random forest method. Int J Geo-Inf 6:365. https://doi.org/10.3390/ijgi6110365

Smith GH (1935) The relative relief of Ohio. Geogr Rev India 25:272–284

Song Y, Gong J, Gao S, Wang D, Cui T, Li Y, Wei B (2012) Susceptibility assessment of earthquake induced landslides using Bayesian network: a case study in Beichuan China. Comput Geosci 42:189–199

Spiker EC, Gori PL (2000) National landslide hazards mitigation strategy: a framework for loss reduction. Department of the interior, U.S. Geol Surv 59

Strahler AN (1952) Hypsometric (area-altitude) analysis of erosional topography. Geol Soc Am Bull 63:117–142

Strahler AN (1964) Quantitative geomorphology of drainage basins and channel networks. In: Chow VT (ed) Handbook of applied hydrology, pp 439–476

Thakkar AK, Dhiman SD (2007) Morphometric analysis and prioritization of mini watershed s in Mohr watersheds. Gujarat using remote sensing and GIS techniques. J Indian Soc Remote Sens 33(1):25–38

Tsangaratos P, Ilia I (2016) Comparison of a logistic regression and Native Bayes classifier in landslide susceptibility assessments: the influence of models complexity and training dataset size. Catena 145:164–179

Umar Z, Pradhan B, Ahmad A et al (2014) Earthquake induced landslide susceptibility mapping using an integrated ensemble frequency ratio and logistic regression models in West Sumatera Province, Indonesia. Catena 118:124–135. https://doi.org/10.1016/j.catena.2014.02.005

Xu C, Dai F, Xu X, Lee YH (2012) GIS-based support vector machine modeling of earthquake-triggered landslide susceptibility in the Jianjiang River watershed China. Geomorphology 145–146:70–80

Yesilnacar EK (2005) The application of computational intelligence to landslide susceptibility mapping in Turkey; Ph.D thesis, Department of Geomatics, University of Melbourne, 423 p

Yin KJ, Yin TZ (1988) Statistical prediction model for slope instability of metamorphosed rock. In: Proceedings of 5th international symposium on landslides Lausanne, Switzerland 2, pp 1269–1272

Chapter 5
Prediction of Landslide Susceptibility Using Bivariate Models

5.1 Introduction

The preparation of landslide susceptibility zonation map is of immense important in site-specific planning and development of Darjeeling Himalaya. Darjeeling Himalaya exhibits an important tourist spots in West Bengal where there is a continuous flow of tourists along with vehicles through mountain roads. The concentration of human population and development of towns such as Darjeeling, Kalimpong, Kurseong and Mirrik in the mountain terrain of Darjeeling Himalaya put enormous pressure over fragile lithology and make the slope more susceptible to landslide activities. The identification of probable landslide locations is very much essential for future planning of the concerned region. Many methods of preparing Landslide Susceptibility and hazard maps have been developed in the last few decades (van Westen et al. 2003; Guzzetti et al. 2005; Lee et al. 2013). Bivariate models like frequency ratio (FR) (Yilmaz 2009; Pradhan and Lee 2010; Yalcin et al. 2011), information value (Info Val) method (van Westen 1997; Xu 2013; Chen et al. 2014), landslide normal risk factor (LNRF) (Gupta and Joshi 1990; Torkashvand et al. 2014), weights of Evidence (WoE) (Pradhan et al. 2010; Mohammady et al. 2012; Pourghasemi et al. 2013), fuzzy Logic (FL) (Suh et al. 2011; Guettouche 2013; Sharma et al. 2013), and statistical index (SI) models (Pourghasemi et al. 2013; Tay et al. 2014; Bourenane et al. 2015), multivariate models like logistic regression (LR) (Nandi and Shakoor 2010; Akgun et al. 2012; Ghosh et al. 2011), artificial neural networks (ANN) (Yilmaz 2009; Pradhan and Lee 2010; Dou et al. 2015) and discriminant analysis (DA) models (Fisher 1936; Davis 2002; Venables and Ripley 2002), pseudo quantitative approach like analytic hierarchy process (AHP) (Yalcin 2008; Ma et al. 2013; Daneshvar 2014).

In India, several statistical techniques were employed to find out different unstable areas for landslides (Mondal and Mandal 2017a, b; Mandal and Mandal 2018). However, there is no general way to prepare Landslide Susceptibility

S. Mandal and S. Mondal, *Geoinformatics and Modelling of Landslide Susceptibility and Risk*, Environmental Science and Engineering,
https://doi.org/10.1007/978-3-030-10495-5_5

(LS) map (Duman et al. 2006; Ma et al. 2013). Every method has its own advantages and disadvantages. So, the LS methods need to be developed for higher precision evaluation (Ma et al. 2013). The main aims of this study was to prepare LS map based on the landslide normal risk factor (LNRF) and statistical index (SI) models in a landslide-prone area located in the Darjeeling Himalaya, India considering 15 landslide CFs such as elevation (El), aspect (As), slope (Sl), curvature (Cu), geology(Ge), soil (So), lineament density (Ld), distance to lineament (D2l), drainage density (Dd), distance to drainage (D2d), stream power index (SPI), topographic wetted index (TWI), rainfall (Ra), normalized differential vegetation index (NDVI) and land use and land cover (LULC) using GIS and remote sensing techniques. SI model was used by Pradhan et al. (2012) for Kulekhani watershed, Nepal; Raman and Punia (2012) for upper Tons river valley, Western Himalaya, India; Mandal and Mandal (2017) for Rorachu river basin of eastern Sikkim Himalaya, India. Similarly LNRF model was applied by Saha et al. (2005) for Garhwal Himalayas; Torkashvand et al. (2014) for Sabalan in the North East of Iran. The present study is associated with the application of four bivariate models i.e. *information value model, landslide nominal risk factor model, Fuzzy logic approach* and *statistical index model* and preparation of landslide susceptibility zonation maps of the Balason river basin of Darjeeling Himalaya.

5.2 Materials and Methods

To identify suitable statistical techniques for preparing landslide susceptibility map of the Balason river basin of Darjeeling Himalaya a comparison has been made between information value model, landslide nominal risk factor model, Fuzzy logic approach and statistical index model with the help of google earth, satellite images, SRTM DEM, topographical maps (Survey of India, SOI), Geological map from Geological survey of India, GSI. All these data were being processed on ARC GIS to prepare data layers of elevation, slope, aspect, curvature, lineament, distance to lineament, drainage density, distance to drainage, geology, geomorphology, soil, land use and land cover, NDVI, stream power index, and topographic wetness index (See Materials and methods section in Chap. 2). Finally several models were applied using ARC GIS 9.3 software to integrate all the data layers to obtain landslide susceptibility maps of the Balason river basin.

5.2.1 Information Value (MIV) Model

Kanungo et al. (2009), Sharma et al. (2009), Vijith et al. (2009), Akbar and Ha (2011), Niu et al. (2011), Pereira et al. (2012), Melo et al. (2012), Balsubramani and Kumaraswamy (2013), Xu (2013), Xu et al. (2013) and recently Chen et al.

(2014), Sharma et al. (2015) successfully used information value model for landslide susceptibility assessment and also used the following equation.

$$I(H, xi) = \log 2 \frac{Npix(xi)/Npix(Ni)}{\sum Npix(xi)/\sum Npix(Ni)} + 1 \quad (5.1)$$

where,
I (H, xi) = the information value of subclass i of a predisposing factor
Npix (xi) = the number of landslide pixels in subclass i
Npix (Ni) = the total pixel number of subclass i
$\sum$Npix(xi) = the total landslide pixel number of in the study area and
$\sum$Npix(Ni) = the total pixel number of the study area.

When no landslides existed in a certain subclass, I (H, xi) equals "0", which would be the smallest value; when I (H, xi) is "1", it means that the ratio of landslide pixels in subclass i equals to the average of the study area; and when the value is larger than "1", more landslides pixels in the subclass lie than those in the study area, and the larger the value, the higher landslides ratio in the subclass.

5.2.2 *Landslide Nominal/Normal/Numerical Risk Factor (LNRF)*

It is one of the most important bivariate statistical model and used by Gupta and Joshi (1990) and Torkashvand et al. (2014) to produce landslide susceptibility map.

$$LNRF_{ij} = \frac{Npix(S_j)}{\left(\sum_{i=1}^{n} Npix(S_j)\right)/n} \quad (5.2)$$

$LNRF_{ij}$ is the Landslide Nominal Risk Factor of the jth class of the ith landslide conditioning factor
$Npix(S_i)$ is the number of pixels containing landslides in jth class
n is the number of classes present in the particular conditioning factor

$$LSI_{LNRF} = \sum_{i=1}^{n} LNRF \quad (5.3)$$

where, *n* is the number of conditioning factors.

"LNRF_elevation" + "LNRF_aspect" + "LNRF_slope" + "LNRF_curvature"+
"LNRF_geology" + "LNRF_soil" + "LNRF_dd" + "LNRF_ld"+
"LNRF_d2l" + "LNRF_d2d" + "LNRF_spi" + "LNRF_twi"+
"LNRF_rainfall" + "LNRF_ndvi" + "LNRF_lulc" (5.4)

5.2.3 Fuzzy Logic Approach for Landslide Susceptibility Assessment

5.2.3.1 Fuzzy Memberships

Membership value is determined by membership function. Membership function is a function that maps all given elements to interval of values <0, 1>.

$$\mu_{\mathrm{A}} : \mathrm{U} \rightarrow \langle 0, 1 \rangle \tag{5.5}$$

where, μA is a membership function, U is a set of elements. Then for each $x \in U$, $\mu A(x)$ is membership value of the element x to the set A (Zadeh 1965). For purpose, we use two functions for computing the fuzzy membership values:

1. Frequency Ratio and
2. Cosine Amplitude.

1. **Frequency Ratio:**
   Frequency Ratio gives proportion of landslide cells in the specific category for each of input layers. It can be described as ratio of relative frequency of landslide cells in a category (an attribute class) to the relative frequency of all landslide cells in the area:

$$\mathrm{FR} = \frac{\mathrm{N\ cell\ (Li)/N\ cell\ (ci)}}{\mathrm{N\ cell\ (L)/N\ cell(C)}} \tag{5.6}$$

   where N cell (Li) is the number of landslide cells in the category *i*, N cell (Ci) is the total number of cells in the category *i*, N cell (L) is total number of landslide cells and N cell (C) is the total number of cells. If the result is higher than 1 it shows higher density of landslide cells in the category then overall in the dataset. Results lower than 1, points to categories that have density of landslide cells lower then density in the dataset. To transform FR to membership values those outputs have to be normalized by dividing each FR by maximal FR in the given group of classes. Then the membership values are from the interval <0, 1> and the higher the number is, the higher is the influence of this category on landslide occurrence.

1. **Cosine Amplitude:**
   Another method for determining the membership values of categories to the set of categories important for landslide occurrence is Cosine Amplitude method:

$$\mathrm{CA} = \frac{\mathrm{N\ cell\ (Li)}}{\sqrt{\mathrm{N\ cell\ (Ci)} \cdot \mathrm{N\ cell\ (L)}}} \tag{5.7}$$

In this case, the membership value is calculated as ratio between number of landslide cells in the category and the square root of its product with the total number of landslide pixels in the dataset. Unlike FR the output values do not have to be normalized because they already fall in interval <0, 1>.

5.2.3.2 Fuzzy Operators

Several fuzzy operators exist for combining membership functions. Best-known operators are AND and OR, but both of them suffer with problem that one of combined sets have significant impact on result of such combination while the other sets do not have such influence. In case of operator AND minimum of all values is the one that defines output and in case of OR operator it is the maximum value. Fuzzy Algebraic Product tends to produce output function lower or equal to the lowest function given, while Fuzzy Algebraic Sum is complementary to the former, so it provides output function higher than all the inputs but never higher than 1. In this study, we use other operator such as Gamma Operation.

Fuzzy Gamma Operation is defined by:

$$\mu_{\gamma} = (\mu_{\text{sum}})\text{k}^{\gamma}.\left(\mu_{\text{product}}\right)^{1-\gamma} \tag{5.8}$$

The exponent γ, which is a number from <0, 1> interval, allows optimization of the membership combination. Setting it to the extremes of the interval give either Fuzzy Algebraic Sum ($\gamma = 1$) or Fuzzy Algebraic Product ($\gamma = 0$).

5.2.3.3 Application of Fuzzy Gamma Operators

We tested several gamma operator (k) values of 0.50, 0.70, 0.80, 0.90, 0.95 and 0.975. It turned that the best performance (AUC) was achieved by $\gamma = 0.975$, making it a parameter of choice for our final susceptibility model. For assigning membership values, Frequency Ratio could be preferred over Cosine Amplitude.

5.2.4 Statistical Index (SI) Model

The statistical index method is associated with the determination of weighted values of a factors class which is defined as the natural algorithm of the landside density of a particular class divided by the landslide density of the study area (Eq. 5.1).

$$wi = \ln\left(\frac{\text{Densclass}}{\text{Densmap}}\right) = \ln\left[\left(\frac{\text{Npix(Si)}}{\text{Npix(Ni)}}\right) \div \left(\frac{\sum \text{Npix(Si)}}{\sum \text{Npix(Ni)}}\right)\right] \tag{5.9}$$

wi is the weight given to the parameter class, ln is the natural logarithm, Densclass is the landslide density within the parameter class, Densmap is the landslide density within the area, Npix (Si) is the number of landslide pixels in parameter class i, Npix (Ni) is the number of pixels in parameter class i, $\sum$ Npix (Si) is the total number of landslide pixels in the area, $\sum$ Npix (Ni) is the total number of pixels in the area.

After the calculation of weights of each class of all parameters, landslide susceptibility index (LSI) map was prepared by the summation of all weighted parameters which can be expressed with the help of following equation:

$$\text{LSI} = \sum_{j=1}^{n} \text{Wij} \tag{5.10}$$

"SI_elevation" + "SI_aspect" + "SI_slope" + "SI_curvature" + "SI_geology" + "SI_soil" + "SI_dd" + "SI_ld" + "SI_d2l" + "SI_d2d" + "SI_spi" + "SI_twi" + "SI_rainfall" + "SI_ndvi" + "SI_LULC"

5.2.5 *Models Validation*

Several techniques have been frequently used for the validation and accuracy assessments of the landslide susceptibility map. Receiver operating characteristics (ROC) curve and frequency ratio plots are the notable ones. Researchers like Akgun et al. (2012) and Tien Bui et al. (2012) employed ROC curve in their studies. It is graphical representation of the sensitivity and 1-specificity of landslide susceptibility models (Mohammady et al. 2012). The value of ROC curve called area under curve value (AUCV) which represents the predictive capability of the landslide susceptibility map. Generally the values of which divided into 5 categories based on the prediction ability i.e. 0.5–0.6 indicates poor, 0.6–0.7 indicates average, 0.7–0.8 indicates good, 0.8–0.9 indicates very good, 0.9–1.0 indicates excellent prediction accuracy (Yesilnacar 2005). Frequency ratio plot is the graphical presentation of the different landslide susceptibility zones against frequency ratio values. It establishes ideal landslide susceptibility map in which the frequency ratio value is continuously increasing from very low to very high landslide susceptibility zones (Pourghasemi et al. 2012 Mondal and Mandal 2017a, b). All the bivariate models were being validated with the help of ROC curve and frequency ratio values (Table 5.1).

Table 5.1 Results of FR of each class of each landslide conditioning factor

Conditioning factor	Class	Number of landslide occurrence pixels	Number of pixel	Modified information value (MIV)	$LNRF_{ij}$	Frequency ratio value	Fuzzy membership value using frequency ratio (FR) method	Weight (wi)
Altitude (m)	106 to 174	0	42,815	0	0.000	0.00	0.00	0
	174 to 267	0	42,258	0	0.000	0.00	0.00	0
	267 to 539	31	41,995	0.277534	0.227	0.23	0.08	−1.48
	539 to 847	271	42,082	1.555519	1.981	1.98	0.71	0.68
	847 to 1065	277	41,913	1.584963	2.025	2.03	0.73	0.71
	1065 to 1257	381	42,012	1.909802	2.785	2.79	1.00	1.03
	1257 to 1432	87	41,959	0.710493	0.636	0.64	0.23	−0.45
	1432 to 1639	28	41,898	0.277534	0.205	0.21	0.08	−1.58
	1639 to 1906	57	41,783	0.510195	0.417	0.42	0.15	−0.87
	1906 to 2603	236	41,785	1.431339	1.725	1.74	0.62	0.55
Aspect	Flat (−1)	4	9005	0.165059	0.029	0.14	0.06	−1.99
	North (0 to 22.5)	2	18,829	0.043069	0.015	0.03	0.01	−3.42
	Northeast (22.5 to 67.5)	14	34,118	0.165059	0.102	0.13	0.06	−2.07
	East (67.5 to 112.5)	53	55,240	0.381871	0.387	0.29	0.13	−1.22
	Southeast (112.5 to 157.5)	134	58,292	0.762961	0.980	0.71	0.31	−0.35
	South (157.5 to 202.5)	619	82,475	1.710493	4.525	2.31	1.00	0.84

(continued)

Table 5.1 (continued)

Conditioning factor	Class	Number of landslide occurrence pixels	Number of pixel	Modified information value (MIV)	$LNRF_{ij}$	Frequency ratio value	Fuzzy membership value using frequency ratio (FR) method	Weight (wi)
	Southwest (202.5 to 247.5)	349	60,795	1.447459	2.551	1.76	0.76	0.57
	West (247.5 to 292.5)	183	56,036	1	1.338	1.00	0.43	0.00
	Northwest (292.5 to 337.5)	9	32,934	0.125531	0.066	0.08	0.03	−2.48
	North (337 5 to 360)	1	12,776	0.043069	0.007	0.02	0.01	−3.73
Slope (°)	0 to 1	7	46,301	0.084889	0.051	0.05	0.01	−3.07
	1 to 4	13	50,128	0.125531	0.095	0.08	0.01	−2.53
	4 to 10	29	40,560	0.277534	0.212	0.22	0.04	−1.52
	10 to 15	94	48,503	0.656046	0.687	0.60	0.11	−0.52
	15 to 19	102	46,043	0.736966	0.746	0.68	0.13	−0.38
	19 to 23	104	46,809	0.736966	0.760	0.68	0.13	−0.38
	23 to 27	136	42,086	0.977974	0.994	0.99	0.18	−0.01
	27 to 31	141	35,846	1.125531	1.031	1.21	0.22	0.19
	31 to 36	172	31,866	1.398549	1.257	1.66	0.31	0.51
	36 to 89	570	32,358	2.662965	4.167	5.41	1.00	1.69
Curvature	Concave	707	193,016	1.084889	1.550	1.13	1.00	0.12
	Flat	57	34,524	0.599462	0.125	0.51	0.45	−0.68
	Convex	604	192,960	0.955606	1.325	0.96	0.85	−0.04

(continued)

Table 5.1 (continued)

Conditioning factor	Class	Number of landslide occurrence pixels	Number of pixel	Modified information value (MIV)	$LNRF_{ij}$	Frequency ratio value	Fuzzy membership value using frequency ratio (FR) method	Weight (wi)
Geology	Darjeeling gneiss	1345	287,309	1.277534	4.916	1.44	1.00	0.36
	Daling series	21	18,518	0.415037	0.077	0.35	0.24	−1.05
	Damuda series/lower Gondwana	0	4784	0	0.000	0.00	0.00	0
	Nahan group	0	16,839	0	0.000	0.00	0.00	0
	Alluvium (older and recent)	2	93,050	0	0.007	0.01	0.01	−5.02
Soil	W002	524	198,569	0.838249	1.915	0.81	0.43	−0.21
	W003	626	102,249	1.510195	2.288	1.88	1.00	0.63
	W004	218	49,222	1.222392	0.797	1.36	0.72	0.31
	W006	0	46,327	0	0.000	0.00	0.00	0
	W009	0	24,133	0	0.000	0.00	0.00	0
Lineament density (Number/km^2)	0 to 0.10	1	39,532	0	0.007	0.01	0.00	−4.86
	0.10 to 0.25	37	43,171	0.347923	0.270	0.26	0.09	−1.33
	0.25 to 0.38	74	43,723	0.599462	0.541	0.52	0.18	−0.65
	0.38 to 0.49	79	43,684	0.628031	0.577	0.56	0.19	−0.59
	0.49 to 0.59	106	41,022	0.838249	0.775	0.80	0.28	−0.23

(continued)

Table 5.1 (continued)

Conditioning factor	Class	Number of landslide occurrence pixels	Number of pixel	Modified information value (MIV)	$LNRF_{ij}$	Frequency ratio value	Fuzzy membership value using frequency ratio (FR) method	Weight (wi)
	0.59 to 0.71	192	43,459	1.222392	1.404	1.36	0.47	0.31
	0.71 to 0.84	177	42,431	1.184425	1.294	1.28	0.44	0.25
	0.84 to 1.00	152	45,012	1.021695	1.111	1.04	0.36	0.04
	1.00 to 1.22	189	39,867	1.277534	1.382	1.46	0.51	0.38
	1.22 to 2.29	361	38,599	1.944291	2.639	2.88	1.00	1.06
Distance from lineament (m)	0 to 10	49	5557	1.874469	0.358	2.71	0.96	1.00
	10 to 20	42	5970	1.642106	0.307	2.16	0.77	0.77
	20 to 40	107	11,709	1.909802	0.782	2.81	1.00	1.03
	40 to 70	154	19,252	1.775785	1.126	2.46	0.88	0.90
	70 to 100	129	19,504	1.584963	0.943	2.03	0.72	0.71
	100 to 200	354	66,656	1.381871	2.588	1.63	0.58	0.49
	200 to 400	239	113,338	0.710493	1.747	0.65	0.23	−0.43
	400 to 700	226	110,616	0.683526	1.652	0.63	0.22	−0.47
	700 to 1200	68	62,986	0.415037	0.497	0.33	0.12	−1.10
	1200 to 1800	0	4912	0	0.000	0.00	0.00	0
Drainage density (m/40,000 m^2)	0 to 53.93	218	119,003	0.628031	1.594	0.56	0.11	−0.57
	53.93 to 143.81	380	97,468	1.125531	2.778	1.20	0.24	0.18
	143.81 to 215.71	274	76,554	1.06413	2.003	1.10	0.22	0.10

(continued)

Table 5.1 (continued)

Conditioning factor	Class	Number of landslide occurrence pixels	Number of pixel	Modified information value (MIV)	$LNRF_{ij}$	Frequency ratio value	Fuzzy membership value using frequency ratio (FR) method	Weight (wi)
	215.71 to 283.12	218	55,176	1.14543	1.594	1.21	0.24	0.19
	283.12 to 355.03	131	37,793	1.043069	0.958	1.07	0.22	0.06
	355.03 to 426.93	39	18,793	0.710493	0.285	0.64	0.13	−0.45
	426.93 to 516.81	39	9641	1.14543	0.285	1.24	0.25	0.22
	516.81 to 620.17	48	4674	2.043069	0.351	3.16	0.64	1.15
	620.17 to 741.51	19	1178	2.555519	0.139	4.96	1.00	1.60
	741.51 to 1145.97	2	220	1.909802	0.015	2.92	0.59	1.03
Distance to drainage (m)	0 to 10	172	41,936	1.165059	1.257	1.26	0.89	0.23
	10 to 20	91	20,857	1.222392	0.665	1.34	0.94	0.29
	20 to 40	214	55,757	1.105353	1.564	1.18	0.83	0.17
	40 to 70	227	65,061	1.043069	1.659	1.07	0.75	0.07
	70 to 100	188	47,405	1.14543	1.374	1.22	0.86	0.20
	100 to 150	247	53,424	1.259387	1.806	1.42	1.00	0.35
	150 to 300	227	69,150	1	1.659	1.01	0.71	0.01

(continued)

Table 5.1 (continued)

Conditioning factor	Class	Number of landslide occurrence pixels	Number of pixel	Modified information value (MIV)	$LNRF_{ij}$	Frequency ratio value	Fuzzy membership value using frequency ratio (FR) method	Weight (wi)
	300 to 700	2	43,921	0	0.015	0.01	0.01	−4.27
	700 to 1500	0	19,226	0	0.000	0.00	0.00	0
	1500 to 2500	0	3763	0	0.000	0.00	0.00	0
SPI	−13.82 to 1.93	10	40,628	0.084889	0.073	0.08	0.04	−2.58
	1.93 to 3.53	37	47,727	0.313158	0.270	0.24	0.12	−1.43
	3.53 to 4.33	157	46,419	1.021695	1.148	1.04	0.53	0.04
	4.33 to 10.07	127	44,636	0.886343	0.928	0.87	0.45	−0.13
	10.07 to 11.00	45	42,730	0.415037	0.329	0.32	0.16	−1.13
	11.00 to 11.67	141	47,117	0.932886	1.031	0.92	0.47	−0.08
	11.67 to 12.34	176	41,514	1.184425	1.287	1.30	0.67	0.26
	12.34 to 13.14	242	38,179	1.540568	1.769	1.95	1.00	0.67
	13.14 to 14.47	238	37,741	1.540568	1.740	1.94	0.99	0.66
	14.47 to 20.21	195	33,809	1.463401	1.425	1.77	0.91	0.57
TWI/CTI	−19.41 to −17.59	248	35,280	1.642106	1.813	2.16	1.00	0.77
	−17.59 to −16.62	63	49,004	0.479168	0.461	0.39	0.18	−0.93
	−16.62 to −14.66	10	42,254	0.084889	0.073	0.07	0.03	−2.62

(continued)

Table 5.1 (continued)

Conditioning factor	Class	Number of landslide occurrence pixels	Number of pixel	Modified information value (MIV)	$LNRF_{ij}$	Frequency ratio value	Fuzzy membership value using frequency ratio (FR) method	Weight (wi)
	−14.66 to −10.75	241	42,590	1.447459	1.762	1.74	0.81	0.55
	−10.75 to −9.92	184	45326	1.165059	1.345	1.25	0.58	0.22
	−9.92 to −9.08	188	44,372	1.184425	1.374	1.30	0.60	0.26
	−9.08 to −8.24	135	40,495	1	0.987	1.03	0.48	0.02
	−8.24 to −7.26	127	40,150	0.977974	0.928	0.98	0.45	−0.03
	−7.26 to −5.87	111	41,303	0.862496	0.811	0.82	0.38	−0.19
	−5.87 to 16.19	61	39,726	0.540568	0.446	0.47	0.22	−0.75
Land use/land cover (LULC)	Agricultural plantation	413	83,710	1.313158	1.208	1.52	0.99	0.42
	Water bodies	147	63,561	0.762961	0.430	0.71	0.46	−0.34
	Settlement	645	128,894	1.330645	1.886	1.54	1.00	0.43
	Forest	163	144,335	0.415037	0.477	0.35	0.23	−1.06

(continued)

Table 5.1 (continued)

Conditioning factor	Class	Number of landslide occurrence pixels	Number of pixel	Modified information value (MIV)	$LNRF_{ij}$	Frequency ratio value	Fuzzy membership value using frequency ratio (FR) method	Weight (wi)
NDVI	−0.14 to 0.09	94	40,463	0.762961	0.687	0.71	0.34	−0.34
	0.09 to 0.16	148	44,788	1	1.082	1.02	0.49	0.02
	0.16 to 0.21	279	41,014	1.613817	2.039	2.10	1.00	0.74
	0.21 to 0.25	277	44,540	1.525461	2.025	1.91	0.91	0.65
	0.25 to 0.28	222	48,426	1.259387	1.623	1.41	0.67	0.34
	0.28 to 0.32	116	41,256	0.886343	0.848	0.87	0.41	−0.15
	0.32 to 0.35	79	41,803	0.656046	0.577	0.58	0.28	−0.54
	0.35 to 0.38	54	35,835	0.540568	0.395	0.46	0.22	−0.77
	0.38 to 0.43	67	45,093	0.540568	0.490	0.45	0.21	−0.78
	0.43 to 0.62	32	37,282	0.347923	0.234	0.26	0.12	−1.33
Rainfall (mm)	2039 to 2560	130	40,635	0.977974	0.950	0.98	0.64	−0.02
	2560 to 2618	209	42,137	1.330645	1.520	1.52	1.00	0.42
	2618 to 2682	186	44,545	1.184425	1.360	1.28	0.84	0.25
	2682 to 2740	161	38,867	1.165059	1.177	1.28	0.84	0.24
	2740 to 2798	186	45,248	1.165059	1.360	1.26	0.83	0.23
	2798 to 2856	147	41,930	1.043069	1.075	1.08	0.71	0.07

(continued)

Table 5.1 (continued)

Conditioning factor	Class	Number of landslide occurrence pixels	Number of pixel	Modified information value (MIV)	$LNRF_{ij}$	Frequency ratio value	Fuzzy membership value using frequency ratio (FR) method	Weight (wi)
	2856 to 2914	87	41,695	0.710493	0.636	0.64	0.42	−0.44
	2914 to 2985	40	42,379	0.347923	0.292	0.29	0.19	−1.24
	2985 to 3100	62	42,956	0.510195	0.453	0.44	0.29	−0.81
	3100 to 3679	161	40,108	1.14543	1.177	1.24	0.82	0.21

Table 5.2 MIV Model, landslide susceptibility zones and frequency ratio (FR)

Landslide susceptibility zones	Total pixels	% of total pixels	Landslide affected pixels	% of landslide affected pixels	Frequency ratio (FR)
Very low	51,395	12.22	0	0.00	0.00
Low	56,329	13.40	51	3.73	0.28
Moderate	108,721	25.86	266	19.48	0.75
High	132,956	31.62	342	24.98	0.79
Very high	71,098	16.91	709	51.82	3.06

5.3 Results and Discussion

5.3.1 *Modified Information Value (MIV) Model and Landslide Susceptibility*

The prepared landslide susceptibility map of the Balason river basin using MIV model was classified into five zones i.e. very low, low, moderate, high and very high susceptibility (Table 5.2 and Fig. 5.1a). The result showed that the Balason river basin dominated by high susceptibility and followed by moderate, very high, low and very low. In very high landslide susceptibility zone, 51.82% pixels are being affected by landslides, in high susceptibility zone 24.98% pixels and in moderate susceptible zone 19.48% pixels are affected by landslides. The estimated frequency ratio of all landslides susceptibility zones revealed that frequency ratio increases from very low to high and very high susceptibility zones. The prepared map based on MIV model may be considered as a tool for site specific management of land and soil for planning and development.

5.3.2 *Landslide Nominal/Normal/Numerical Risk Factor (LNRF) and Landslide Susceptibility*

The application of LNRF model and prepared landslide susceptibility zonation map of the Balason river basin is classified into five zones i.e. very low, low, moderate, high and very high landslide susceptibility (Fig. 5.1b). The Balason river basin is dominated by high landslide susceptibility and followed by moderate, very high low and very low susceptibility. In the Balson river basin, 51.77% area of very high landslide susceptibility zone is affected by landslides and 26.35% area of high susceptibility zone is affected by landslides. The derived frequency ratio of very low, low, moderate, high and very high landslide susceptibility zones are 0.00, 0.24, 0.68, 0.86 and 3.47 respectively (Table 5.3).

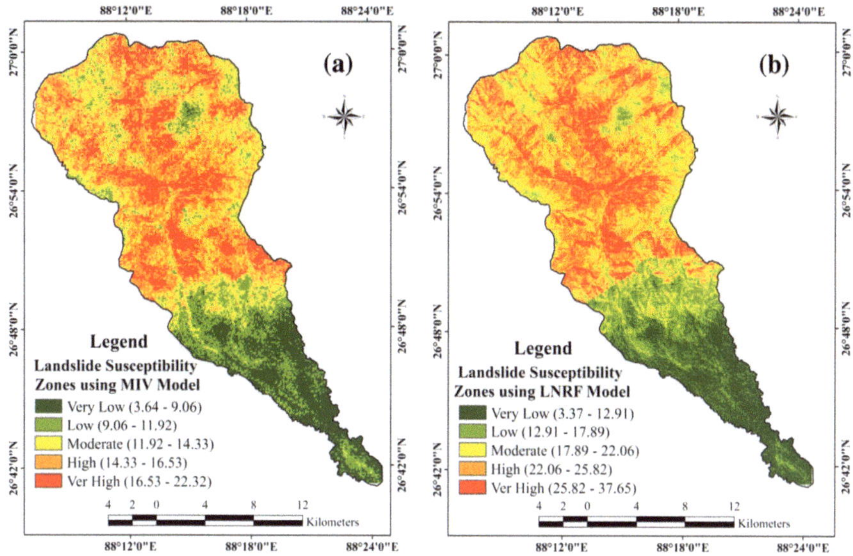

Fig. 5.1 Landslide susceptibility maps **a** MIV model and **b** LNRF model

Table 5.3 LNRF model, landslide susceptibility zones and frequency ratio (FR)

Landslide susceptibility zones	Total pixels	% of total pixels	Landslide affected pixels	% of landslide affected pixels	Frequency ratio (FR)
Very low	56,810	13.51	0	0.00	0.00
Low	57,793	13.74	46	3.35	0.24
Moderate	114,045	27.12	253	18.52	0.68
High	129,046	30.69	361	26.35	0.86
Very high	62,806	14.94	708	51.77	3.47

5.3.3 *Fuzzy Logic (FL) and Landslide Susceptibility*

The developed landslide susceptibility zonation map of the Balason river basin of Darjeeling Himalaya using fuzzy logic approach was grouped into five i.e. very low, low, moderate, high and very high landslide susceptibility (Fig. 5.2c). The study denotes that only 7.17 and 13.93% areas are covered by high and very high landslide susceptibility. In very high and high landslide susceptibility zones, 54.38 and 14.98% areas are affected by landslides. The frequency ratio value increases from very low to very high landslide susceptibility (Table 5.4). High FR value is being found in very high landslide susceptible areas of the Balason river basin.

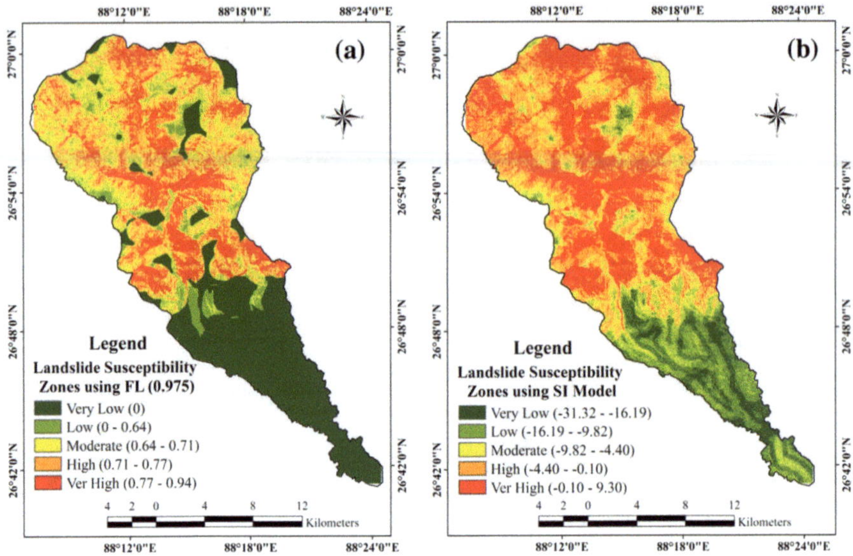

Fig. 5.2 Landslide susceptibility maps **c** fuzzy logic approach and **d** SI model

Table 5.4 FL model, landslide susceptibility zones and frequency ratio (FR)

Landslide susceptibility zones	Total pixels	% of total pixels	Landslide affected pixels	% of landslide affected pixels	Frequency ratio (FR)
Very low	127,349	30.29	31	2.27	0.07
Low	98,332	23.38	137	10.00	0.43
Moderate	106,112	25.23	251	18.37	0.73
High	30,143	7.17	205	14.98	2.09
Very high	58,564	13.93	744	54.38	3.90

5.3.4 Statistical Index (SI) Model and Landslide Susceptibility

SI model and prepared landslide susceptibility map of the Balason river basin has been classified into five zones i.e. very low, low, moderate, high and very high landslide susceptibility (Fig. 5.2d). More than 75% area of the Balason river basin is experienced by moderate to very high landslide susceptibility (Table 5.5). The study also revealed that 95% landslide affected pixels are being observed in moderate, high and very high landslide susceptibility zones. The estimated FR value is very low in

Table 5.5 SI Model, landslide susceptibility zones and frequency ratio (FR)

Landslide susceptibility zones	Total pixels	% of total pixels	Landslide affected pixels	% of landslide affected pixels	Frequency ratio (FR)
Very low	29,540	7.02	0	0.00	0.00
Low	54,472	12.95	6	0.46	0.04
Moderate	74,514	17.72	140	10.20	0.58
High	148,865	35.40	319	23.30	0.66
Very high	113,109	26.90	903	66.04	2.46

moderate and low landslide susceptible areas. High FR values in very high landslide susceptibility zone is found which demonstrated high landslide susceptibility is accompanied with high probability of landslide phenomena in the Balason river basin.

5.4 Validation Results

In the present study, receiver operating characteristics (ROC) curve and frequency ratio values were being considered for the validation of the prepared landslide susceptibility maps using information value mode, landslide nominal risk factor model, fuzzy logic approach and statistical index model. The area under curve (AUC) value of the prepared LSI maps were 0.744 or 74% for information value mode (IVM), 0.742 or 74% for landslide nominal risk factor model (LNRFM), 0.975 or 97% for fuzzy logic approach (FLA) and 0.797 or 79% for statistical index model (SIM). The result of ROC curve depicts that fuzzy logic approach has high % accuracy (97%) and may be accepted as one of the most significant bivariate model in landslide susceptibility studies of the Balson river basin of Darjeeling Himalaya (Fig. 5.3). The statistical index model ranks second with 79% accuracy and followed by landslide nominal risk factor model and information value model.

5.5 Conclusion

All the prepared landslide susceptibility zonation maps of the Balson river basin of Darjeeling Himalaya depicted that the lower most section of the basin having low landslide occurrences phenomena and low probability and upper and middle most segments are very much prone to landslide phenomena. The middle most section of a slope unit is attributed with more hydrostatic pressure as a result of drainage concentration over the space. The Balason river basin is an elongated one which helps in

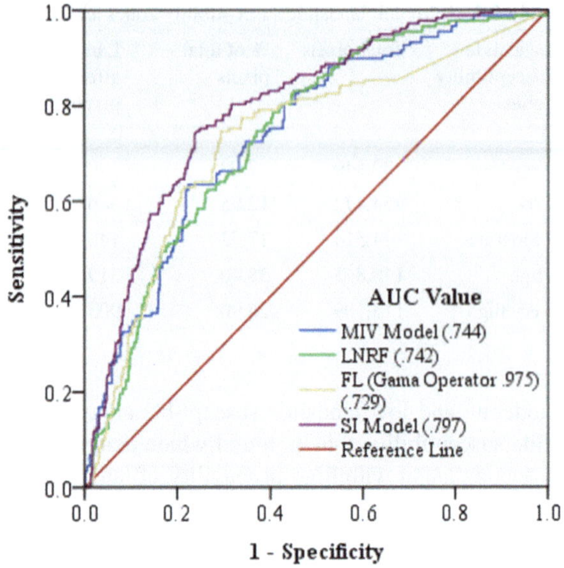

Fig. 5.3 ROC curve of MIV mode, LNRF model, fuzzy logic and SI model

accumulation of surface water and promote slope saturation, soil erosion, and mass movement. Fuzzy logic approach shows high level of accuracy i.e. 97.5% and this very approach may be taken into account in landslide susceptibility studies. The study in regard to landslide susceptibility requires a proper understanding of various factors i.e. slope, geology, soil, drainage, and human intervention. At present, expansion of human settlement and associated activities over the steep mountain slope inviting destructive landslide hazards. The planner and developers should follow a landslide susceptibility zonation map for the expansion of human settlement and construction of roads in the mountain terrain which may check the destructive character of landslide events. So, the application of statistical techniques contribute a lot in the preparation of landslide susceptibility map and planning as well as development of the mountain environment of Darjeeling Himalaya.

References

Akbar T, Ha S (2011) Landslide hazard zoning along Himalayan Kaghan Valley of Pakistan—by integration of GPS, GIS, and remote sensing technology. Landslide 8(4):527–540. https://doi.org/10.1007/s10346-011-0260-1

Akgun A, Sezer EA, Nefeslioglu HA, Gokceoglu C, Pradhan B (2012) An 277 easy-to-use MATLAB program (MamLand) for the assessment of landslide susceptibility using a Mamdani fuzzy algorithm. Comput Geosci 38:23–34

Balsubramani K, Kumaraswamy K (2013) Application of geospatial technology and information value technique in landslide hazard zonation mapping: a case study of Giri Valley, Himachal Pradesh. Disaster Adv 6:38–47

Bourenane H, Bouhadad Y, Guettouche MS, Braham M (2015) GIS based landslide susceptibility zonation using bivariate statistical and expert approaches in the city of Constantine northeast Algeria. Bull Eng Geol Environ 74(2):337–355

Chen W, Li W, Hou E, Zhao Z, Deng N, Bai H, Wang D (2014) Landslide susceptibility mapping based on GIS and information value model for the Chencang district of Baoji, China. Arab J Geosci. https://doi.org/10.1007/s12517-014-1369-z

Daneshvar MRM (2014) Landslide susceptibility zonation using analytical hierarchy process and GIS for the Bojnurd region, northeast of Iran. Landslides 11(6):1079–1091

Davis JC (2002) Statistics and data analysis in geology, 3rd edn. Wiley, p 638; ERDAS (1997) ERDAS field guide, 5th edn. ERDAS Inc., Atlanta, Georgia, USA, p 672

Dou J, Bui DT, Yunus AP, Jia K, Song X, Revhaug I, Xia H, Zhu Z (2015) Optimization of causative factors for landslide susceptibility evaluation using remote sensing and GIS data in parts of Niigata, Japan. PloS ONE 10(7):e0133262

Duman TY, Can T, Gokceoglu C, Nefeslioglu HA, Sonmez H (2006) Application of logistic regression for landslide susceptibility zoning of Cekmece area, Istanbul, Turkey. Environ Geol 51(2):241–256

Fisher RA (1936) The use of multiple measurements in taxonomic problems. Ann Eugenics 7:179–188

Ghosh S, Carranza EJM, van Westen CJ, Jetten VG, Bhattacharya DN (2011) Selecting and weighting spatial predictors for empirical modeling of landslide susceptibility in the Darjeeling Himalayas (India). Geomorphology 131(1):35–56

Guettouche MS (2013) Modeling and risk assessment of landslides using fuzzy logic: application on the slopes of the Algerian Tell (Algeria). Arab J Geosci 6:3163–3173

Gupta RP, Joshi BC (1990) Landslide hazard zonation using the GIS approach—a case study from the Ramganga Catchment, Himalayas. Eng Geol 28:119–131

Guzzetti F, Reichenbach P, Cardinali M, Galli M, Ardizzone F (2005) Probabilistic landslide hazard assessment at the basin scale. Geomorphology 72(1):272–299

Kanungo D, Arrora M, Sarkar S, Gupta R (2009) Landslide susceptibility zonation (LSZ) mapping—a review. J South Asia Disaster Stud 2:81–105

Lee S, Hwang J, Park I (2013) Application of data-driven evidential belief functions to landslide susceptibility mapping in Jinbu, Korea. Catena 100:15–30

Ma F, Wang J, Yuan R, Zhao H, Guo J (2013) Application of analytical hierarchy process and least-squares method for landslide susceptibility assessment along the Zhong-Wu natural gas pipeline, China. Landslides 10(4):481–492

Mohammady M, Pourghasemi HR, Pradhan B (2012) Landslide susceptibility mapping at Golestan Province, Iran: a comparison between frequency ratio, Dempster-Shafer, and weights-of-evidence models. J Asian Earth Sci 61:221–236

Mondal S, Mandal S (2017a) RS and GIS-based landslide susceptibility mapping of the Balason River basin, Darjeeling Himalaya, using logistic regression (LR) model. Georisk Assess Manag Risk Eng Syst Geohazards 12:1, 29–44. https://doi.org/10.1080/17499518.2017.1347949

Mondal S, Mandal S (2017b) Application of frequency ratio (FR) model in spatial prediction of landslides in the Balason river basin, Darjeeling Himalaya. Spat Inf Res. https://doi.org/10.1007/s41324-017-0101-y

Mandal S, Mandal K (2017) Bivariate statistical index for landslide susceptibility mapping in the Rorachu River basin of Eastern Sikkim Himalaya, India. Spat Inf Res. https://doi.org/10.1007/s41324-017-0156-9

Mandal S, Mandal K (2018) Modeling and mapping landslide susceptibility zones using GIS based multivariate binary logistic regression (LR) model in the Rorachu river basin of eastern Sikkim Himalaya, India. Model Earth Syst Environ. https://doi.org/10.1007/s40808-018-0426-0

Melo R, Vieira G, Caselli A, Ramos M (2012) Susceptibility modelling of hummocky terrain distribution using the information value method (Deception Island, Antarctic Peninsula). Geomorphology 155, 156:88–95

Nandi A, Shakoor A (2010) A GIS-based landslide susceptibility evaluation using bivariate and multivariate statistical analyses. Eng Geol 110(1):11–20

Niu QF, Cheng WM, Lan HX, Liu Y, Xie YW (2011) Susceptibility assessment of secondary geological disaster based on information value methodology for Yushu earthquake region (In Chinese). J Mountain Sci 29:243–249

Pereira S, Zezere J, Bateira C (2012) Assessing predictive capacity and conditional independence of landslide predisposing factors for shallow landslide susceptibility models. Nat Hazards Earth Syst Sci 12:979–988. https://doi.org/10.5194/nhess129792012

Pourghasemi HR, Mohammady M, Pradhan M (2012) Landslide susceptibility mapping using index of entropy and conditional probability models in GIS: Safarood Basin, Iran. Catena 97:71–84

Pourghasemi HR, Goli Jirandeh A, Pradhan B, Xu C, Gokceoglu C (2013) Landslide susceptibility mapping using support vector machine and GIS. J Earth Syst Sci 122(2):349–369

Pradhan B, Lee S (2010) Landslide susceptibility assessment and factor effect analysis: backpropagation artificial neural networks and their comparison with frequency ratio and bivariate logistic regression modelling. Environ Model Softw 25(6):747–759

Pradhan B, Oh HJ, Buchroithner M (2010) Weights-of-evidence model applied to landslide susceptibility mapping in a tropical hilly area. Geomat Nat Hazards Risk 1(3):199–223

Pradhan MSA, Dawadi A, Kim T (2012) Use of different bivariate statistical landslide susceptibility methods: a case study of Kulekhani watershed, Nepal. J Nepal Geol Soc 44(2012):1–12

Saha AK, Gupta RP, Sarkar I, Arora KM, Csaplovics E (2005) An approach for GIS-based statistical landslide susceptibility zonation with a case study in the Himalayas. Landslides 2(1):61–69

Sharma L, Patel N, Ghosh M, Debnath P (2009) Geographical information system based landslide probabilistic model with trivariate approach—A case study in Sikkim Himalaya. In: Eighteenth united nations regional cartographic conference for Asia and the Pacific, UN Economic and Social Council, Bankok

Sharma LP, Nilanchal Patel, Ghose MK, Debnath P (2013) Synergistic application of fuzzy logic and geoinformatics for landslide vulnerability zonation—a case study in Sikkim Himalayas, India. Appl Geomat 5:271–284

Sharma LP, Patel N, Ghose MK, Debnath P (2015) Development and application of Shannon's entropy integrated information value model for landslide susceptibility assessment and zonation in Sikkim Himalaya in India. Nat Hazards 75:1555–1576

Suh J, Choi Y, Roh TD, Lee HJ, Park HD (2011) National-scale assessment of landslide susceptibility to rank the vulnerability to failure of rock-cut slopes along expressways in Korea. Environ Earth Sci 63(3):619–632

Tay LT, Lateh H, Hossain MK, Kamil AA (2014) Landslide hazard mapping using a Poisson distribution: a case study in Penang Island, Malaysia. In Landslide science for a safer geoenvironment. Springer International Publishing, pp 521–525

Tien Bui D, Pradhan B, Lofman O, Revhaug I (2012) Landslide susceptibility assessment in Vietnam using support vector machines, decision tree and Naïve Bayes models. Math Probl Eng. https://doi.org/10.1155/2012/9746382012:26

Torkashvand AM, Irani A, Sorur J (2014) The preparation of landslide map by landslide numerical risk factor (LNRF) model and geographic information system (GIS). Egypt J Remote Sens Space Sci 17:159–170

van Westen CJ (1997) Statistical landslide hazard analysis. In: Application guide, ILWIS 2.1 for Windows. ITC, Enschede, Netherlands, pp 73–84

van Westen CJ, Rengers N, Soeters R (2003) Use of geomorphological information in indirect landslide susceptibility assessment. Nat Hazards 30(3):399–419

Venables WN, Ripley BD (2002) Modern applied statistics with S, 4th edn. Springer, Berlin, Germany, p 495

Vijith H, Rejith PG, Madhu G (2009) Using Info Val method and GIS techniques for the spatial modelling of landslides susceptibility in the upper catchment of river Meenachil in Kerala. J Indian Soc Remote Sens 37:241–250

Xu C (2013) Assessment of earthquake-triggered landslide susceptibility based on expert knowledge and information value methods: a case study of the 20 April 2013 Lushan, China Mw6. 6 earthquake. Disaster Adv, 6(13):119–130

Xu WB, Yu WJ, Jing SC, Zhang GP, Huang JX (2013) Debris flow susceptibility assessment by GIS and information value model in a large-scale region, Sichuan Province (China). Nat Hazards 65:1379–1392

Yalcin A (2008) GIS-based landslide susceptibility mapping using analytical hierarchy process and bivariate statistics in Ardesen (Turkey): comparisons of results and confirmations. Catena 72:1–12

Yalcin A, Reis S, Aydinoglu AC, Yomralioglu T (2011) A GIS based comparative study of frequency ratio, analytical hierarchy process, bivariate statistics and logistics regression methods for landslide susceptibility mapping in Trabzon, NE Turkey. Catena 85(3):274–287

Yesilnacar EK (2005) The application of computational intelligence to landslide susceptibility mapping in Turkey. Ph.D. thesis, Department of Geomatics, University of Melbourne, p 423

Yilmaz I (2009) Landslide susceptibility mapping using frequency ratio, logistic regression, artificial neural networks and their comparison: a case study from Kat landslides (Tokat Turkey). Comput Geosci 35(6):1125–1138

Zadeh LA (1965) Fuzzy sets. Information and control, vol 8/3. Elsevier, pp 338–353

Chapter 6
Probabilistic Approaches and Landslide Susceptibility

6.1 Introduction

Landslides are one of the destructive events in any terrestrial environment especially in mountainous area. Losses occur when people and their cultural structures are come in contact to landslides. The degree of losses and damages depend on the exposure, intensity and vulnerability of landslides. According to the International disaster database, it account about 4.9% of all natural disasters and 54% occurred in Asia (Guha-Sapir et al. 2018). In addition, worldwide it took about 1000 people lives annually and damages approximately 4 billion US property damages (Lee and Pradhan 2007). In India, landslide occurrences mostly concentrated in the Himalayan mountain regions, the North-eastern hill ranges, the Western Ghats, the Nilgiris, the Eastern Ghats and the Vindhya region. Himalayas alone count for landslides of every fame, name and description—big and small, quick and creeping, ancient and new. The occurrences of landslides in Darjeeling Himalayan mountain region are attributed of several geo-environmental conditions like topographic, hydrologic, geologic and anthropogenic along with triggering factors. In recent times the process of slope movement has been accelerated due to extensive deforestation, unsystematic and unscientific constructional activities, inadequate drainage system, urban expansion, rapid changes of land use practice etc. It causes great damages to properties and lives in every year especially in rainy season. Several studies on landslides of Darjeeling Himalaya stated about the landslide damages (Basu and Sarkar 1987; Sengupta 1995; Kanungo et al. 2006; Mandal and Maiti 2014). About 72 people, 127 people and 667 people were died due to the landslides events in the year of 1899, 1950 and 1968 respectively and Darjeeling, Ghum and Kurseong town were badly affected by Jan, 1934 landslide events (Biswas and Pal 2015). Beside these, several landslide disasters occurred in the year of 1980, 1991, 1993, 2003, 2005, 2009 and 2015 in Darjeeling Himalaya (Starkel and Basu 2000; Sarkar et al. 2010). Keeping in view the facts of landslide damages, it is necessary to prepare landslide susceptibility map of Darjeeling Himalaya. Various scientific approaches have been applied to mapping

S. Mandal and S. Mondal, *Geoinformatics and Modelling of Landslide Susceptibility and Risk*, Environmental Science and Engineering,
https://doi.org/10.1007/978-3-030-10495-5_6

landslide susceptibility zones with the help of geo-informatics. van Westen et al. (2006) divided landslide susceptibility models into two approaches. First one was associated with the deterministic and physical mechanism which discusses the geo-technical properties of rocks and soils. Second one was related with the location of previous landslides and landslide causative factors which leads to the development of bivariate, multivariate and probabilistic models. Bivariate models like frequency ratio (FR) (Pradhan and Lee 2010; Yalcin et al. 2011), information value (Info Val) method (van Westen 1997; Chen et al. 2014), landslide normal risk factor (LNRF) (Gupta and Joshi 1990; Torkashvand et al. 2014), weights of Evidence (WOE) (Mohammady et al. 2012; Pourghasemi et al. 2013a), fuzzy Logic (FL) (Guettouche 2013; Sharma et al. 2013), and statistical index (SI) models (Pourghasemi et al. 2013b; Bourenane et al. 2015) were frequently incorporated in landslide susceptibility mapping. Several multivariate techniques like logistic regression (LR) (Akgun et al. 2012; Ghosh et al. 2011) and discriminant analysis (DA) models (Davis 2002; Venables and Ripley 2002), and probabilistic models like bayesian probability (BP) (Lee et al. 2002; Sujatha et al. 2014) and certainty factor (CF) approach (Pradhan and Kim 2014; Wang et al. 2015) used in different parts of the world. In addition, some machine learning algorithms (MLAs) like Bayes-based algorithms such as Naïve Bayes (Bui et al. 2012; Pourghasemi and Rahmati 2017) and Bayes Networks (Pham et al. 2016); functional algorithms such as Artificial Neural Networks (ANN) (Park et al. 2013; Tsangaratos and Benardos 2014), Support Vector Machine (SVM) (Peng et al. 2014; Hong et al. 2015), Decision Tree (DT) algorithms (Pradhan 2013; Naghibi et al. 2016) etc. were employed by landslide researchers. However, no generalization has been made for landslide susceptibility modelling till now. In this study, data-driven bivariate statistical weights of evidence (WOE) model was used which has the ability to calculate statistically the weight values of each class of causative factor which reduce the dependency on the subjective knowledge, to use incomplete data coverage in input maps and to avoid correlated/dependent causative factors etc. Lee et al. (2002), van Westen et al. (2003), Lee and Choi (2004), Thiery et al. (2004, 2007), Mathew et al. (2007), Neuhäuser and Terhorst (2007), Poli and Sterlacchini (2007), Rezaei Moghaddam et al. (2007), Dahal et al. (2008a, b), Regmi et al. (2014) applied WOE model in different section of the world and generated reasonable landslide susceptibility maps. The main aim of this study is to prepare landslide susceptibility map based on *certainty factor model* and *weight of evidence model* considering 15 landslide causative factors such as elevation, aspect, slope, curvature, geology, soil, lineament density, distance to lineament, drainage density, distance to drainage, stream power index (SPI), topographic wetted index (TWI), rainfall (Ra), normalized differential vegetation index (NDVI) and land use and land cover (LULC) for the Darjeeling Himalaya.

6.2 Materials and Methods

6.2.1 Weights of Evidence (WoE) Model

The data-driven bivariate statistical weights of evidence (WOE) model is basically a Bayesian approach in log-linear version used to calculate probability using the concept of prior and posterior probability. This approach is based on the information generated from the relationship between landslide distribution and landslide causative factors. WOE model was first applied in medical purpose (Spiegelhater and Knill-Jones 1984), then in mineral potential mapping (Bonham-Carter et al. 1988, 1989; Bonham-Carter 1994) and recently used in landslide susceptibility mapping (Dahal et al. 2008a, b; Regmi et al. 2014). The weight based on the WOE model was calculated using the following formula as suggested by Bonham-Carter et al. (1989).

$$W_i^+ = \ln\frac{P\{B|S\}}{P\{B|\overline{S}\}} \tag{6.1}$$

$$W_i^- = \ln\frac{P\{\overline{B}|S\}}{P\{\overline{B}|\overline{S}\}} \tag{6.2}$$

where, W_i^+ is the weights of evidence when the factor B is present, W_i^- is the weights of evidence when the factor B is absent, P is the probability, B is the presence of landslide predictive factor, $\overline{B}$ is the absence of a landslide predictive factor, S is the presence of a landslide, and $\overline{S}$ is the absence of a landslide. Based on Eqs. (6.3) and (6.4), the weights of evidence can be rewritten in pixels numbers (van Westen et al. 2003).

$$W_i^+ = \ln\frac{\frac{Npix_1}{Npix_1+Npix_2}}{\frac{Npix_3}{Npix_3+Npix_4}} \tag{6.3}$$

$$W_i^- = \ln\frac{\frac{Npix_2}{Npix_1+Npix_2}}{\frac{Npix_4}{Npix_3+Npix_4}} \tag{6.4}$$

where, $Npix_1$ is the number of landslide pixels present in a given factor class, $Npix_2$ is the number of landslide pixels not present in the same given factor class, $Npix_3$ is the number of pixels in the given factor class in which no landslide pixels are present and $Npix_4$ is the number of pixels in the given factor class when neither a landslide nor the given factor is present.

For each factor, W_i^+ is used to indicate the importance of the presence of the factor for the landslide occurrences. Positive and negative W_i^+ value means the presence of the factor is favourable for the occurrence of landslides and not favourable for the occurrence of landslides respectively. W_i^- is used to represent the importance of the absence of the factor for the occurrence of landslides. Positive W_i^- value indicates the absence of the factor is favourable for the occurrence of landslides, and negative for vice versa. Extreme weights value is useful for the landslide susceptibility mapping; zero weights value represents no relation with the occurrence of landslide (Table 6.1). The difference between the W_i^+ and W_i^- weights is known as the weights contrast, C ($C = W_i^+ + W_i^-$), where C reflects the overall relation between a predictable variable and landslide occurrence. A contrast value equal to zero indicates that no significant role in susceptibility analysis. A positive contrast value represents a positive spatial association with landslide susceptibility and vice versa for a negative contrast value (Corsini et al. 2009).

The landslide susceptibility index (LSI) map using WOE model was prepared by summing the contrast value of each factor.

$$\text{LSI} = \Sigma \text{C} \tag{6.5}$$

where, C = contrast of each factor.

Equation can be rewritten as:

$$\begin{aligned}\text{LSI} = {} & \text{“WOE\_Elevation”} + \text{“WOE\_Aspect”} + \text{“WOE\_Slope”} \\ & + \text{“WOE\_Curvature”} + \text{“WOE\_Geology”} + \text{“WOE\_Soil”} \\ & + \text{“WOE\_Lineament density”} + \text{“WOE\_Distance to lineamnet”} \\ & + \text{“WOE\_Drainage density”} + \text{“WOE\_Distance to drainage”} \\ & + \text{“WOE\_SPI”} + \text{“WOE\_TWI”} + \text{“WOE\_Rainfall”} \\ & + \text{“WOE\_NDVI”} + \text{“WOE\_LULC”}\end{aligned} \tag{6.6}$$

6.2.2 *Certainty Factor (CF) Model*

Certainty factor approach is dealt with the problem of combination of non-homogeneous data. This model has been successfully used by Kanungo et al. (2011), Sujatha et al. (2012) and Liu et al. (2014) in landslide susceptibility mapping studies.

$$\text{CF} = \begin{cases} \frac{ppa - pps}{ppa(1 - pps)} & if\ ppa \geq pps \\ \frac{ppa - pps}{ppa(1 - ppa)} & if\ ppa < pps \end{cases} \tag{6.7}$$

where CF is the certainty factor (Table 6.1), ppa is the conditional probability of having a number of landslides in a class ‘a’ and pps is the prior probability of having the total number of landslides in the study area ‘A’.

The layers are combined pairwise according to the integration rules (Chung and Fabbri 1993; Binaghi et al. 1998). The combination of CF values of two thematic layers ‘z’ is expressed in the following equation as given by Binaghi et al. (1998).

Table 6.1 Results of FR of each class of each landslide conditioning factor

Conditioning factor	Class	Number of pixel	Number of landslide occurrence pixels	W_i^+	W_i^-	C	Number of landslide events	ppa	pps	CF
Altitude (m)	106–174	42,815	0	−4.086	0.1078	−4.1938	0	0.000000	0.00049	−1.00
	174–267	42,258	0	−4.086	0.1063	−4.1923	0	0.000000	0.00049	−1.00
	267–539	41,995	31	−1.4857	0.0826	−1.5683	20	0.000476	0.00049	−0.03
	539–847	42,082	271	0.6860	−0.1157	0.8017	36	0.000855	0.00049	0.43
	847–1065	41,913	277	0.7121	−0.1216	0.8338	34	0.000811	0.00049	0.40
	1065–1257	42,012	381	1.0310	−0.2218	1.2529	19	0.000452	0.00049	−0.08
	1257–1432	41,959	87	−0.4516	0.0395	−0.4912	17	0.000405	0.00049	−0.17
	1432–1639	41,898	28	−1.5853	0.0846	−1.6699	20	0.000477	0.00049	−0.03
	1639–1906	41,783	57	−0.8710	0.0623	−0.9333	19	0.000455	0.00049	−0.07
	1906–2603	41,785	236	0.5540	−0.0850	0.6390	41	0.000981	0.00049	0.50
Aspect	Flat (−1)	9005	4	−1.9940	0.0188	−2.0127	2	0.000222	0.00049	−0.55
	North (0–22.5)	18,829	2	−3.4251	0.0445	−3.4696	5	0.000266	0.00049	−0.46
	Northeast (22.5–67.5)	34,118	14	−2.0733	0.0746	−2.1479	8	0.000234	0.00049	−0.52
	East (67.5–112.5)	55,240	53	−1.2234	0.1017	−1.3250	15	0.000272	0.00049	−0.45
	Southeast (112.5–157.5)	58,292	134	−0.3482	0.0463	−0.3945	45	0.000772	0.00049	0.37
	South (157.5–202.5)	82,475	619	0.8402	−0.3851	1.2253	64	0.000776	0.00049	0.37

(continued)

Table 6.1 (continued)

Conditioning factor	Class	Number of pixel	Number of landslide occurrence pixels	W_i^+	W_i^-	C	Number of landslide events	ppa	pps	CF
	Southwest (202.5–247.5)	60,795	349	0.5704	−0.1388	0.7092	38	0.000625	0.00049	0.22
	West (247.5–292.5)	56,036	183	0.0038	−0.0006	0.0044	22	0.000393	0.00049	−0.20
	Northwest (292.5–337.5)	32,934	9	−2.4799	0.0752	−2.5551	6	0.000182	0.00049	−0.63
	North (337.5–360)	12,776	1	−3.7304	0.0302	−3.7606	1	0.000078	0.00049	−0.84
Slope (°)	0–1	46,301	7	−3.0720	0.1119	−3.1839	1	0.000022	0.00049	−0.96
	1–4	50,128	13	−2.5323	0.1178	−2.6501	2	0.000040	0.00049	−0.92
	4–10	40,560	29	−1.5177	0.0803	−1.5980	10	0.000247	0.00049	−0.50
	10–15	48,503	94	−0.5193	0.0515	−0.5709	21	0.000433	0.00049	−0.12
	15–19	46,043	102	−0.3853	0.0386	−0.4239	24	0.000521	0.00049	0.06
	19–23	46,809	104	−0.3824	0.0391	−0.4215	32	0.000684	0.00049	0.28
	23–27	42,086	136	−0.0067	0.0007	−0.0075	34	0.000808	0.00049	0.39
	27–31	35,846	141	0.1905	−0.0197	0.2103	27	0.000753	0.00049	0.35
	31–36	31,866	172	0.5084	−0.0557	0.5642	24	0.000753	0.00049	0.35
	36–89	32,358	570	1.7036	−0.4601	2.1638	31	0.000958	0.00049	0.49
Curvature	Concave	193,016	707	0.1190	−0.1133	0.2323	91	0.000471	0.00049	−0.04
	Flat	34,524	57	−0.6799	0.0433	−0.7231	21	0.000608	0.00049	0.19
	Convex	192,960	604	−0.0387	0.0317	−0.0704	94	0.000487	0.00049	−0.01

(continued)

Table 6.1 (continued)

Conditioning factor	Class	Number of pixel	Number of landslide occurrence pixels	W_i^+	W_i^-	C	Number of landslide events	ppa	pps	CF
Geology	Darjeeling Gneiss	287,309	1345	0.3654	−2.9390	3.3044	194	0.000675	0.00049	0.27
	Daling Series	18,518	21	−1.0560	0.0297	−1.0857	6	0.000324	0.00049	−0.34
	Damuda Series/Lower Gondwana	4784	0	−4.086	0.0115	−4.0975	0	0.000000	0.00049	−1.00
	Nahan Group	16,839	0	−4.086	0.0410	−4.1270	0	0.000000	0.00049	−1.00
	Alluvium (older and recent)	93,050	2	−5.0229	0.2496	−5.2725	6	0.000064	0.00049	−0.87
Soil	W002	198,569	524	−0.2099	0.1567	−0.3666	103	0.000519	0.00049	0.06
	W003	102,249	626	0.6352	−0.3341	0.9692	79	0.000773	0.00049	0.37
	W004	49,222	218	0.3097	−0.0493	0.3589	24	0.000488	0.00049	0.00
	W006	46,327	0	−4.086	0.1171	−4.2031	0	0.000000	0.00049	−1.00
	W009	24,133	0	−4.086	0.0593	−4.1453	0	0.000000	0.00049	−1.00
Lineament density (Number/km^2)	0–0.10	39,532	1	−4.8600	0.0983	−4.9583	10	0.000253	0.00049	−0.48
	0.10–0.25	43,171	37	−1.3363	0.0812	−1.4175	8	0.000185	0.00049	−0.62
	0.25–0.38	43,723	74	−0.6550	0.0544	−0.7094	15	0.000343	0.00049	−0.30
	0.38–0.49	43,684	79	−0.5886	0.0504	−0.6390	18	0.000412	0.00049	−0.16
	0.49–0.59	41,022	106	−0.2310	0.0221	−0.2531	17	0.000414	0.00049	−0.15
	0.59–0.71	43,459	192	0.3072	−0.0423	0.3495	31	0.000713	0.00049	0.31

(continued)

Table 6.1 (continued)

Conditioning factor	Class	Number of pixel	Number of landslide occurrence pixels	W_i^+	W_i^-	C	Number of landslide events	ppa	pps	CF
	0.71–0.84	42,431	177	0.2495	−0.0323	0.2818	36	0.000848	0.00049	0.42
	0.84–1.00	45,012	152	0.0374	−0.0046	0.0420	16	0.000355	0.00049	−0.27
	1.00–1.22	39,867	189	0.3780	−0.0492	0.4273	29	0.000727	0.00049	0.33
	1.22–2.29	38,599	361	1.0621	−0.2107	1.2728	26	0.000674	0.00049	0.27
Distance from lineament (m)	0–10	5557	49	1.0027	−0.0232	1.0259	10	0.001800	0.00049	0.73
	10–20	5970	42	0.7751	−0.0169	0.7920	6	0.001005	0.00049	0.51
	20–40	11,709	107	1.0387	−0.0534	1.0921	13	0.001110	0.00049	0.56
	40–70	19,252	154	0.9044	−0.0728	0.9772	18	0.000935	0.00049	0.48
	70–100	19,504	129	0.7129	−0.0517	0.7646	18	0.000923	0.00049	0.47
	100–200	66,656	354	0.4922	−0.1272	0.6194	38	0.000570	0.00049	0.14
	200–400	113,338	239	−0.4347	0.1225	−0.5572	42	0.000371	0.00049	−0.24
	400–700	110,616	226	−0.4664	0.1251	−0.5915	44	0.000398	0.00049	−0.19
	700–1200	62,986	68	−1.1052	0.1117	−1.2169	17	0.000270	0.00049	−0.45
	1200–1800	4912	0	−4.086	0.0118	−4.0978	0	0.000000	0.00049	−1.00
Drainage density (m/40,000 m^2)	0–53.93	119,003	218	−0.5757	0.1597	−0.7354	14	0.000118	0.00049	−0.76
	53.93–143.81	97,468	380	0.1816	−0.0619	0.2436	33	0.000339	0.00049	−0.31
	143.81–215.71	76,554	274	0.0958	−0.0226	0.1184	43	0.000562	0.00049	0.13
	215.71–283.12	55,176	218	0.1950	−0.0330	0.2280	54	0.000979	0.00049	0.50
	283.12–355.03	37,793	131	0.0636	−0.0065	0.0701	28	0.000741	0.00049	0.34
	355.03–426.93	18,793	39	−0.4508	0.0169	−0.4676	18	0.000958	0.00049	0.49

(continued)

Table 6.1 (continued)

Conditioning factor	Class	Number of pixel	Number of landslide occurrence pixels	W_i^+	W_i^-	C	Number of landslide events	ppa	pps	CF
	426.93–516.81	9641	39	0.2187	−0.0057	0.2244	12	0.001,245	0.00049	0.61
	516.81–620.17	4674	48	1.1566	−0.0246	1.1812	1	0.000214	0.00049	−0.56
	620.17–741.51	1178	19	1.6140	−0.0112	1.6252	2	0.001698	0.00049	0.71
	741.51–1145.97	220	2	1.0335	−0.0009	1.0344	1	0.004545	0.00049	0.89
Distance to drainage (m)	0–10	41,936	172	0.2325	−0.0294	0.2619	37	0.000882	0.00049	0.44
	10–20	20,857	91	0.2946	−0.0180	0.3126	23	0.001103	0.00049	0.56
	20–40	55,757	214	0.1659	−0.0280	0.1939	4	0.000072	0.00049	−0.85
	40–70	65,061	227	0.0702	−0.0134	0.0836	43	0.000661	0.00049	0.26
	70–100	47,405	188	0.1988	−0.0283	0.2271	42	0.000886	0.00049	0.45
	100–150	53,424	247	0.3528	−0.0635	0.4163	35	0.000655	0.00049	0.25
	150–300	69,150	227	0.0090	−0.0018	0.0108	20	0.000289	0.00049	−0.41
	300–700	43,921	2	−4.2721	0.1092	−4.3813	2	0.000046	0.00049	−0.91
	700–1500	19,226	0	−4.086	0.0470	−4.1330	0	0.000000	0.00049	−1.00
	1500–2500	3763	0	−4.086	0.0090	−4.0950	0	0.000000	0.00049	−1.00
SPI	−13.82 to 1.93	40,628	10	−2.5845	0.0946	−2.6791	2	0.000049	0.00049	−0.90
	1.93–3.53	47,727	37	−1.4367	0.0934	−1.5301	13	0.000272	0.00049	−0.44
	3.53–4.33	46,419	157	0.0390	−0.0049	0.0440	37	0.000797	0.00049	0.39
	4.33–10.07	44,636	127	−0.1344	0.0148	−0.1493	14	0.000314	0.00049	−0.36
	10.07–11.00	42,730	45	−1.1301	0.0740	−1.2041	15	0.000351	0.00049	−0.28
	11.00–11.67	47,117	141	−0.0838	0.0101	−0.0939	25	0.000531	0.00049	0.08

(continued)

Table 6.1 (continued)

Conditioning factor	Class	Number of pixel	Number of landslide occurrence pixels	W_i^+	W_i^-	C	Number of landslide events	ppa	pps	CF
	11.67–12.34	41,514	176	0.2658	−0.0339	0.2997	28	0.000674	0.00049	0.27
	12.34–13.14	38,179	242	0.6701	−0.0998	0.7699	29	0.000760	0.00049	0.36
	13.14–14.47	37,741	238	0.6649	−0.0974	0.7623	21	0.000556	0.00049	0.12
	14.47–20.21	33,809	195	0.5751	−0.0702	0.6453	22	0.000651	0.00049	0.25
TWI/CTI	−19.41 to −17.59	35,280	248	0.7742	−0.1127	0.8870	33	0.000935	0.00049	0.48
	−17.59 to −16.62	49,004	63	−0.9304	0.0770	−1.0074	24	0.000490	0.00049	0.00
	−16.62 to −14.66	42,254	10	−2.6238	0.0989	−2.7227	3	0.000071	0.00049	−0.86
	−14.66 to −10.75	42,590	241	0.5559	−0.0873	0.6432	25	0.000587	0.00049	0.17
	−10.75 to −9.92	45,326	184	0.2222	−0.0305	0.2527	27	0.000596	0.00049	0.18
	−9.92 to −9.08	44,372	188	0.2652	−0.0364	0.3016	25	0.000563	0.00049	0.13
	−9.08 to −8.24	40,495	135	0.0245	−0.0026	0.0272	22	0.000543	0.00049	0.10
	−8.24 to −7.26	40,150	127	−0.0282	0.0029	−0.0311	14	0.000349	0.00049	−0.29
	−7.26 to −5.87	41,303	111	−0.1916	0.0188	−0.2105	20	0.000484	0.00049	−0.01

(continued)

Table 6.1 (continued)

Conditioning factor	Class	Number of pixel	Number of landslide occurrence pixels	W_i^+	W_i^-	C	Number of landslide events	ppa	pps	CF
	−5.87 to 16.19	39,726	61	−0.7525	0.0538	−0.8063	13	0.000327	0.00049	−0.33
Land use/Land cover (LULC)	Agricultural plantation	83,710	413	0.4181	−0.1378	0.5559	52	0.000621	0.00049	0.21
	Water bodies	63,561	147	−0.3422	0.0504	−0.3925	17	0.000267	0.00049	−0.45
	Settlement	128,894	645	0.4324	−0.2724	0.7048	59	0.000458	0.00049	−0.07
	Forest	144,335	163	−1.0602	0.2947	−1.3549	78	0.000540	0.00049	0.09
NDVI	−0.14–0.09	40,463	94	−0.3377	0.0301	−0.3678	9	0.000222	0.00049	−0.55
	0.09–0.16	44,788	148	0.0157	−0.0019	0.0175	14	0.000313	0.00049	−0.36
	0.16–0.21	41,014	279	0.7412	−0.1258	0.8671	23	0.000561	0.00049	0.13
	0.21–0.25	44,540	277	0.6509	−0.1146	0.7656	22	0.000494	0.00049	0.01
	0.25–0.28	48,426	222	0.3443	−0.0549	0.3992	24	0.000496	0.00049	0.01
	0.28–0.32	41,256	116	−0.1463	0.0147	−0.1610	22	0.000533	0.00049	0.08
	0.32–0.35	41,803	79	−0.5445	0.0454	−0.5899	25	0.000598	0.00049	0.18
	0.35–0.38	35,835	54	−0.7714	0.0490	−0.8203	18	0.000502	0.00049	0.02
	0.38–0.43	45,093	67	−0.7855	0.0634	−0.8489	27	0.000599	0.00049	0.18
	0.43–0.62	37,282	32	−1.3348	0.0694	−1.4042	22	0.000590	0.00049	0.17
Rainfall (mm)	2039–2560	40,635	130	−0.0168	0.0018	−0.0186	34	0.000837	0.00049	0.41

(continued)

Table 6.1 (continued)

Conditioning factor	Class	Number of pixel	Number of landslide occurrence pixels	W_i^+	W_i^-	C	Number of landslide events	ppa	pps	CF
	2560–2618	42,137	208	0.4186	−0.0595	0.4782	19	0.000451	0.00049	−0.08
	2618–2682	44,545	186	0.2505	−0.0343	0.2848	37	0.000831	0.00049	0.41
	2682–2740	38,867	161	0.2425	−0.0283	0.2708	30	0.000772	0.00049	0.37
	2740–2798	45,248	186	0.2348	−0.0324	0.2672	16	0.000354	0.00049	−0.28
	2798–2856	41,930	147	0.0750	−0.0087	0.0837	20	0.000477	0.00049	−0.03
	2856–2914	41,695	87	−0.4453	0.0388	−0.4841	17	0.000408	0.00049	−0.17
	2914–2985	42,379	40	−1.2397	0.0768	−1.3166	11	0.000260	0.00049	−0.47
	2985–3100	42,956	62	−0.8145	0.0616	−0.8761	6	0.000140	0.00049	−0.71
	3100–3679	40,108	161	0.2109	−0.0251	0.2360	16	0.000235	0.00049	−0.19

$$Z = \begin{cases} x + y - xy, \ x, \ y \geq 0 \\ \frac{x+y}{1-\min(|x|,|y|)} \ x, y \text{ opposite sign} \\ x + y + xy, \ x, y < 0 \end{cases} \tag{6.8}$$

6.2.3 *Models Validation*

Akgun et al. (2012), Bui et al. (2012) and Ozdemir and Altural (2013) applied ROC curve in validating landslide susceptibility assessment studies. It is graphical representation of the sensitivity and 1-specificity of landslide susceptibility models (Mohammady et al. 2012). The value of ROC curve called area under curve value (AUCV) which represents the predictive capability of the landslide susceptibility map.

6.3 Weights of Evidence (WOE) Model, Landslide Susceptibility Zones and Frequency Ratio

Weight of evidence model based landslide susceptibility zonation map of the Balason river basin has been classified into five i.e. very low, low, moderate, high and very high landslide susceptibility (Fig. 6.1a). More than 65% area of the basin is experienced with high landslide susceptibility (Table 6.2) and remaining area is existing under moderate to very low landslide susceptibility. In very high landslide susceptibility zones of the Balason river basin more than 65% area is affected by landslide processes and 25.27% area is affected by landslide activities in high landslide susceptibility zones. The estimated frequency ratio value of all classes of landslide susceptibility denotes that very high landslide susceptibility zone having high probability of landslide phenomena and vice versa.

Table 6.2 Weight of evidence model, landslide susceptibility and frequency ratio

Landslide Susceptibility Zones	Total pixels	% of total pixels	Landslide affected pixels	% of landslide affected pixels	Frequency ratio (FR)
Very low	39,242	9.33	0	0.00	0.00
Low	50,077	11.91	1	0.10	0.01
Moderate	53,904	12.82	106	7.77	0.61
High	148,826	35.39	346	25.27	0.71
Very high	128,451	30.55	915	66.86	2.19

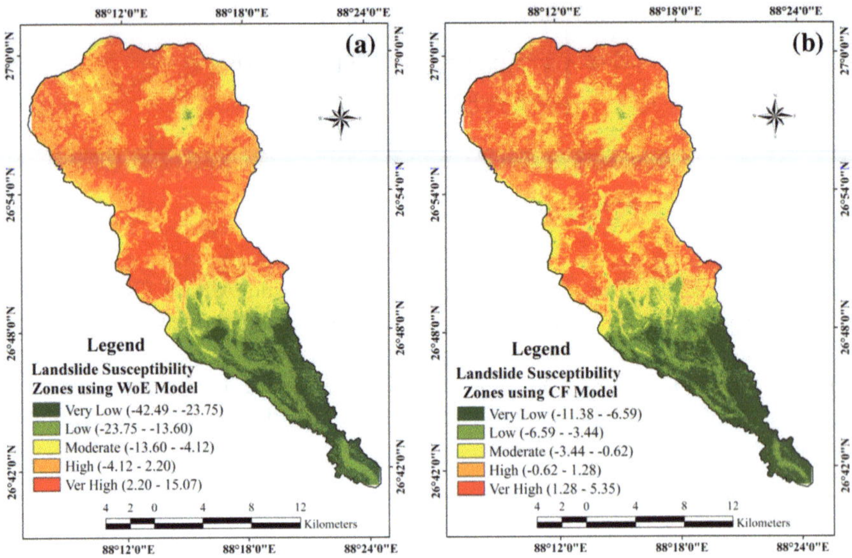

Fig. 6.1 Landslide susceptibility zonation map of the Balason river basin. **a** Weight of evidence model (WoEM) and **b** Certainty factor model (CFM)

6.4 Certainty Factor (CF) Model, Landslide Susceptibility Zones and Frequency Ratio

The CFM induced landslide susceptibility zonation map of the Balason river basin is dominated by high landslide susceptibility (33.93%) which is followed by very high, moderate, low and very low landslide susceptibility (Fig. 6.1b). Very high and high landslide susceptibility zones are highly affected by landslides where 50% area is affected by landslides in very high susceptibility zones (Table 6.3). On the other hand low and very low susceptibility zones are least affected by landslide phenomena. It can be assumed that CFM has provided a result which can be accepted a best suited landslide model of Darjeeling Himalaya. The moderate, high and very high landslide susceptibility zones are registered with frequency ratio value of 0.34, 1.30 and 1.88 respectively.

6.5 Validation Results

The area under curve (AUC) value of the prepared LSI maps were 0.756 or 75% for weight of evidence model (WoEM) and 0.849 or 84% for certainty factor model (CFM). The result of ROC curve depicts that certainty factor model (CFM) has high

Table 6.3 Certainty factor model, landslide susceptibility and frequency ratio

Landslide susceptibility zones	Total pixels	% of total pixels	Landslide affected pixels	% of landslide affected pixels	Frequency ratio (FR)
Very low	52,135	12.40	0	0.00	0.00
Low	42,571	10.12	1	0.10	0.01
Moderate	71,506	17.00	78	5.71	0.34
High	142,662	33.93	605	44.24	1.30
Very high	111,627	26.55	683	49.95	1.88

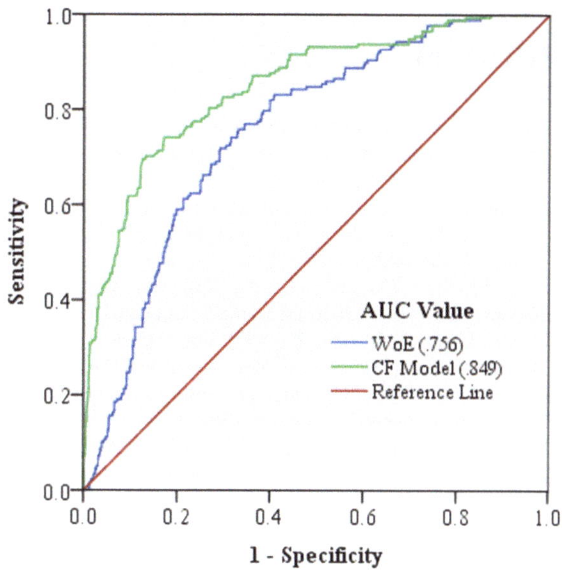

Fig. 6.2 ROC curve of WoE mode and certainty factor model

% accuracy (84%) and may be accepted as one of the most significant bivariate model in landslide susceptibility studies of the Balason river basin of Darjeeling Himalaya (Fig. 6.2).

6.6 Conclusion

The study has made a comparison between weight of evidence model (WoEM) and certainty factor model (CFM). Both models were being applied to prepare landslide susceptibility zonation map of the Balason river basin of Darjeeling Himalaya, India integrating landslide conditioning factors of slope, elevation, aspect, curvature, geology, geomorphology, soil, lineament density, distance to lineament, drainage density, distance to drainage, stream power index, topographic wetness index, Land use and land cover and NDVI. The relative significance of all these factors were being studied in consultation with landslide locations of the Balason river basin and accordingly class weighted values and certainty factor values were estimated and landslide susceptibility zonation maps were being produced. The study demonstrated that CF model is more reliable than WoE Model because of high % of accuracy level. Finally, it could be assessed that CF model can be used for preparing landslide susceptibility zonation map of any geomorphic unit over the earth surface applying expert's knowledge domain with regard to landslides and causative factors for site specific mountain slope management.

References

Akgun A, Sezer EA, Nefeslioglu HA, Gokceoglu C, Pradhan B (2012) An easy-to-use MATLAB program (MamLand) for the assessment of landslide susceptibility using a Mamdani fuzzy algorithm. Comput Geosci 38:23–34

Basu SR, Sarkar S (1987) Ecosystem vis-a vis landslide: a case study in Darjeeling Himalaya, proce. Impact of development on environment. Geog Soc India 2, 45–53. Unpublished GSI report, FS 1981–82, 1982–83

Biswas SS, Pal R (2015) Causes of Landslides in Darjeeling Himalayas during June-July, 2015. J Geogr Nat Disasters 2016(6):2. https://doi.org/10.4172/2167-0587.1000173

Binaghi E, Luzi L, Madella P (1998) Slope instability zonation: a comparison between certainty factor and fuzzy Dempster–Shafer approaches. Nat Hazards 17:77–97

Bonham-Carter GF (1994) Geographic information systems for geoscientists: modelling with GIS. In: Computer methods in the geosciences, vol 13. Pergamon Press, Oxford, p 398

Bonham-Carter GF, Agterberg FP, Wright DF (1988) Integration of geological datasets for gold exploration in Nova Scotia. Photogramm Eng Remote Sens 54:1585–1592

Bonham-Carter, G.F., Agterberg, F.P. and Wright, D.F., (1989) Weights of evidence modelling: a new approach to mapping mineral potential. In: Statistical applications in the earth science. Geological survey of Canada, Paper 89–9, 171–183

Bourenane H, Bouhadad Y, Guettouche MS, Braham M (2015) GIS based landslide susceptibility zonation using bivariate statistical and expert approaches in the city of Constantine northeast Algeria. Bull Eng Geol Environ 74(2):337–355

Bui DT, Pradhan B, Lofman O, Revhaug I, Dick OB (2012) Spatial prediction of landslide hazards in Hoa Binh province (Vietnam): a comparative assessment of the efficacy of evidential belief functions and fuzzy logic models. Catena 96:28–40

Chen W, Li W, Hou E, Zhao Z, Deng N, Bai H, Wang D (2014) Landslide susceptibility mapping based on GIS and information value model for the Chencang District of Baoji, China, 2014. Arab J Geosci. https://doi.org/10.1007/s12517-014-1369-z

Chung CF, Fabbri AG (1993) The representation of geosciences information for data integration. Nonrenew Resour 2(2):122–139

Corsini A, Cervi F, Daehne A, Ronchetti F (2009) Coupling geomorphic field observation and LIDAR derivatives to map complex landslides. In: Malet JP, Remaître A, Bogaard T (eds) Landslides processes—from geomorphologic mapping to dynamic modeling, proceedings of the landslide processes conference, 6–7 February 2009, Strasbourg

Dahal RK, Hasegawa S, Nonomura A, Yamanaka M, Dhakal S, Paudyal P (2008a) Predictive modelling of rainfall-induced landslide hazard in the Lesser Himalaya of Nepal based on weights-of-evidence. Geomorphology 102:496–510

Dahal RK, Hasegawa S, Nonomura A, Yamanaka M, Masuda T, Nishino K (2008b) GIS-based weights-of evidence modelling of rainfall-induced landslides in small catchments for landslide susceptibility mapping. Environ Geol 54(2):314–324

Davis JC (2002) Statistics and data analysis in geology, 3rd ed. Wiley, 638 pp. ERDAS, 1997. ERDAS field guide, 5th edn. ERDAS, Inc., Atlanta, Georgia, USA, 672 pp

Ghosh S, Carranza EJM, van Westen CJ, Jetten VG, Bhattacharya DN (2011) Selecting and weighting spatial predictors for empirical modeling of landslide susceptibility in the Darjeeling Himalayas (India). Geomorphology 131(1):35–56

Guettouche MS (2013) Modeling and risk assessment of landslides using fuzzy logic: application on the slopes of the Algerian Tell (Algeria). Arab J Geosci 6:3163–3173

Guha-Sapir D, Below R, Hoyois PH (2018) EM-DAT: international disaster database. http://www.emdat.be, Université Catholique de Louvain, Brussels, Belgium, last access 19 Feb 2018

Gupta RP, Joshi BC (1990) Landslide hazard zonation using the GIS approach—a case study from the Ramganga Catchment, Himalayas. Eng Geol 28:119–131

Hong H, Pradhan B, Xu C, Bui DT (2015) Spatial prediction of landslide hazard at the Yihuang area (China) using two-class kernel logistic regression, alternating decision tree and support vector machines. CATENA 133:266–281

Kanungo DP, Arora MK, Sarkar S, Gupta RP (2006) A comparative study of conventional, ANN black box, fuzzy and combined neural and fuzzy weighting procedures for landslide susceptibility Zonation in Darjeeling Himalayas. Eng Geol 85:347–366

Kanungo DP, Sarkar S, Sharma S (2011) Combining Neural Network with fuzzy, certainty factor and likelihood ratio concepts for spatial prediction of landslide. Nat Hazards 59(3):1491–1512

Lee S, Choi J (2004) Landslide susceptibility mapping using GIS and the weight-of-evidence model. Int J Geogr Inf Sci 18(8):789–814. https://doi.org/10.1080/13658810410001702003

Lee S, Choi J, Min K (2002) Landslide susceptibility analysis and verification using the Bayesian probability model. Environ Geol 43(1–2):120–131

Lee S, Pradhan B (2007) Landslide hazard mapping at Selangor Malaysia using frequency ratio and logistic regression models. Landslides 4:33–41

Liu M, Chen X, Yang S (2014) Collapse landslide and mudslide hazard zonation. In: Landslide science for a safer geoenvironment. Springer International Publishing, pp 457–462

Mandal S, Maiti R (2014) Role of lithological composition and lineaments in landsliding: a case study of Shivkhola watershed. Darjeeling Himal Int J Geol Earth Environ Sci 4(1):126–132

Mathew J, Jha VK, Rawat GS (2007) Weights of evidence modelling for landslide hazard zonation mapping in part of Bhagirathi valley, Uttarakhand. Curr Sci 92(5):628–638

Mohammady M, Pourghasemi HR, Pradhan B (2012) Landslide susceptibility mapping at Golestan Province, Iran: a comparison between frequency ratio, Dempster-Shafer, and weights-of-evidence models. J Asian Earth Sci 61:221–236

Naghibi SA, Pourghasemi HR, Dixon B (2016) GIS-based groundwater potential mapping using boosted regression tree, classification and regression tree, and random forest machine learning models in Iran. Environ Monit Assess 188(1):1–27

Neuhäuser B, Terhorst B (2007) Landslide susceptibility assessment using "weights-of-evidence" applied to a study area at the Jurassic escarpment (SW-Germany). Geomorphology, 86(1-2):12–24. https://doi.org/10.1016/j.geomorph.2006.08.002

Ozdemir A, Altural T (2013) A comparative study of frequency ratio, weights of evidence and logistic regression methods for landslide susceptibility mapping: Sultan Mountains, SW Turkey. J Asian Earth Sci 64:180–197

Park S, Choi C, Kim B, Kim J (2013) Landslide susceptibility mapping using frequency ratio, analytic hierarchy process, logistic regression, and artificial neural network methods at the Inje area, Korea. Environ Earth Sci 68:1443–1464

Peng L, Niu R, Huang B, Wu X, Zhao Y, Ye R (2014) Landslide susceptibility mapping based on rough set theory and support vector machines: a case of the Three Gorges area, China. Geomorphology 204:287–301

Pham BT, Tien Bui D, Pham HV (2016) Spatial prediction of rainfall induced landslides using Bayesian Network at Luc Yen District, Yen Bai Province (Viet Nam). In: International conference on environmental issues in mining and natural resources development (EMNR 2016), Hanoi University of mining and geology (HUMG), Viet Nam, pp 1–10

Poli S, Sterlacchini S (2007) Landslide representation strategies in susceptibility studies using weights of-evidence modeling technique. Nat Resour Res 16:121–134

Pourghasemi HR, Pradhan B, Gokceoglu C, Mohammadi M, Moradi HR (2013a) Application of weights-of-evidence and certainty factor models and their comparison in landslide susceptibility mapping at Haraz watershed, Iran. Arabian J Geosci 6(7):2351–2365

Pourghasemi HR, Moradi HR, Aghda SF (2013b) Landslide susceptibility mapping by binary logistic regression, analytical hierarchy process and statistical index models and assessment of their performances. Nat Hazards 69(1):749–779

Pourghasemi HR, Rahmati O (2017) Prediction of the landslide susceptibility: which algorithm, which precision?

Pradhan AMS, Kim YT (2014) Relative effect method of landslide susceptibility zonation in weathered granite soil: a case study in Deokjeok-ri Creek, South Korea. Nat Hazards 72(2):1189–1217

Pradhan B, Lee S (2010) Landslide susceptibility assessment and factor effect analysis: backpropagation artificial neural networks and their comparison with frequency ratio and bivariate logistic regression modelling. Environ Model Softw 25(6):747–759

Pradhan B (2013) A comparative study on the predictive ability of the decision tree, support vector machine and neuro-fuzzy models in landslide susceptibility mapping using GIS. Comput Geosci 51:350–365

Regmi AD, Devkota KC, Yoshida K, Pradhan B, Pourghasemi HR, Kumamoto T, Akgun A (2014) Application of frequency ratio, statistical index, and weights-of-evidence models and their comparison in landslide susceptibility mapping in Central Nepal Himalaya. Arab J Geosci 7:725–742

Rezaei Moghaddam MH, Khayyam M, Ahmadi M, Farajzadeh M (2007) Mapping susceptibility landslide by using the weight–of-evidence model: a case study in Merek valley. Iran. J Appl Sci 7(22):3342–3355

Sarkar S et al (2010) Geo-hazards in sub Himalayan North Bengal. Department of Geography & Applied Geography, University of North Bengal, West Bengal, p 2010

Sengupta CK (1995) Detailed study of geofactors in selected hazard prone stretches along the surface communication routes in parts of Darjeeling Himalaya. Unpublished GSI Report, FS 1993–94 & 1994–95

Sharma LP, Nilanchal P, Ghose MK, Debnath P (2013) Synergistic application of fuzzy logic and geoinformatics for landslide vulnerability zonation—a case study in Sikkim Himalayas, India. Appl Geomat 5:271–284

Spiegelhater D, Knill-Jones RP (1984) Statistical and knowledge approaches to clinical decision-support systems, with an application in gastroenterology. J R Stat Soc 147:35–77

Starkel L, Basu SR (2000) Landslides and floods in the Darjiling Himalayas. New Delhi, 1–168: Indian Science Academy

Sujatha ER, Kumaravel P, Rajamanickam GV (2014) Assessing landslide susceptibility using Bayesian probability-based weight of evidence model. Bull Eng Geol Environ 73:147. https://doi.org/10.1007/s10064-013-0537-9

Sujatha ER, Rajamanikam GV, Kumaravel P (2012) Landslide susceptibility analysis using probabilistic certainty factor approach: a case study on Tevankarai stream watershed, India. J Earth Syst Sci 121(5):1337–1350

Thiery Y, Malet JP, Sterlacchini S, Puissant A, Maquaire O (2007) Landslide susceptibility assessment by bivariate methods at large scales: application to a complex mountainous environment. Geomorphology 92:38–59

Thiery Y, Sterlacchini S, Malet JP, Puissant A, Remaître A, Maquaire O (2004) Strategy to reduce subjectivity in landslide susceptibility zonation by GIS in complex mountainous environments. In: Toppen F, Prastacos P (eds) Proceedings of AGILE 2004: 7th AGILE conference on geographic information science. 29th Apr–1st May 2004, Heraklion, Greece, pp 623–634

Torkashvand AM, Irani A, Sorur J (2014) The preparation of landslide map by Landslide Numerical Risk Factor (LNRF) model and Geographic Information System (GIS). Egypt J Remote Sens Space Sci 17:159–170

Tsangaratos P, Benardos A (2014) Estimating landslide susceptibility through an artificial neural network classifier. Nat Hazards 74(3):1–28

van Westen CJ (1997) Statistical landslide hazard analysis. In: Application guide, ILWIS 2.1 for Windows. ITC, Enschede, The Netherlands, pp 73–84

van Westen CJ, Rengers N, Soeters R (2003) Use of geomorphological information in indirect landslide susceptibility assessment. Nat Hazards 30:399–419

van Westen CJ, van Asch TWJ, Soeters R (2006) Landslide hazard and risk zonation: why is it still so difficult? Bull Eng Geol Environ 65(5):167–184

Venables WN, Ripley BD (2002) Modern applied statistics with S, 4th edn. Springer, Berlin, Germany, p 495

Wang Q, Li W, Chen W, Bai H (2015) GIS-based assessment of landslide susceptibility using certainty factor and index of entropy models for the Qianyang County of Baoji city, China. J Earth Syst Sci 124(7):1399–1415

Yalcin A, Reis S, Aydinoglu AC, Yomralioglu T (2011) A GIS-based comparative study of frequency ratio, analytical hierarchy process, bivariate statistics and logistics regression methods for landslide susceptibility mapping in Trabzon, NE Turkey. CATENA 85(3):274–287

Chapter 7
Machine Learning Models and Spatial Distribution of Landslide Susceptibility

7.1 Introduction

To develop landslide susceptibility distribution map of the Balason river basin of Darjeeling Himalaya artificial neural network (ANN) model has been applied using geospatial tools. *Artificial neural network model* was successfully applied by Xie et al. (2004), Pistocchi et al. (2002), Lee et al. (2003, 2004) and Lee (2007), Pradhan (2010a, b, c) for landslide susceptibility assessment and prediction. The present study identified various landslide causative factors existing in the Balason river basin i.e. slope, aspect, curvature, geology, geomorphology, soil, distance to drainage, drainage density, distance to lineament, lineament density, TWI, SPI, LULC, and rainfall. All these data layers were being integrated with the help of support vector machine (SVM) and *artificial neural network model* in consultation with geospatial tools to generate landslide susceptibility zonation map of the corresponding study area.

The main objectives of the present study is to identify landslide susceptibility zones with the of Artificial Neural Networks (ANNs), which is associated with black box models to be implemented, similar to several other statistical approaches (Carrara et al. 1991, 1995). In the present day, ANN is used for various scientific and engineering applications (Emami et al. 1998; Caparrini et al. 1996) which starts a database variables and describes the particular process. ANN is applied to assess indirect determination of the triggering parameters (Mayoraz et al. 1996) as well as to assess landslide susceptibility mapping, with physical terrain factors (Lee et al. 2001). Many mathematical techniques such as decision tree algorithm (Hong et al. 2005; Tsangaratos and Ilia 2016; Wu et al. 2014); fuzzy logic algorithm (Ilanloo 2011; Zhu et al. 2014; Bui et al. 2012); neuro-fuzzy algorithm (Sezer et al. 2011; Bui et al. 2012) have been introduced for evaluating landslide susceptible areas all over the world.

In the present day, machine learning algorithms are being successfully used for developing quantitative models as well for assessing spatial distribution of landslide susceptibility. Ada and San (2018) applied and compared different machine

S. Mandal and S. Mondal, *Geoinformatics and Modelling of Landslide Susceptibility and Risk*, Environmental Science and Engineering,
https://doi.org/10.1007/978-3-030-10495-5_7

learning methods such as *support vector machines* (SVM) and *random forest* (RF) for developing landslide susceptibility assessment and prediction model and preparing landslide susceptibility zonation map. Other machine learning methods are *Fisher's Linear Discriminant Analysis* (FLDA), *Bayesian Network* (BN), *Logistic Regression* (LR), and *Naïve Bayes* (NB). Pham et al. (2016) used all these machine learning models in landslide susceptibility studies. Goetz et al. (2015) compared several machine learning algorithms including Random Forest (RF), SVM, and LR and analysed for predicting landslides. Yet, there are so many machine learning methods in landslide susceptibility assessment and prediction, the introduction of new sophisticated techniques and algorithms are still required for developing the high accuracy based landslide susceptibility assessment methods.

7.2 Supprot Vector Machine (SVM) and Artificial Neural Network (ANN) Models: Materials and Methods

7.2.1 Support Vector Machine (SVM)

The supervised learning method, SVM is based on the theory of statistical learning and the principle of structural risk minimization (Vapnik 1995). SVM basically try to find out an optimal separating hyperplane which differentiate between the two classes (i.e., landslides and no landslides), and the set of training data (Pradhan 2013). For discriminant-type statistical problems, SVM modelling includes hyperplane that separates the data patterns and the use of kernel functions for the conversion of the original nonlinear data patterns (Yao et al. 2008).

Researchers like Yao et al. (2008); Xu et al. (2012a, b) described the two-class SVM modelling in detail. Consider a set of linear separable training vectors x_i ($i = 1, 2, \ldots, n$). The training vectors consist of two classes, denoted as $y_i = \pm 1$. The aim of the SVM is to find out an n-dimensional hyperplane which differentiate the two classes by their maximum gap.

$$\frac{1}{2}||w||^2 \quad (7.1)$$

Subject to the following constraints:

$$y_i((w.x_i) + b) \geq 1 \quad (7.2)$$

where, $||w||$ is the norm of the hyperplane, b is a scalar base, and (.) denotes the scalar product operation. Using the Lagrangian multiplier, the cost function can be defined as:

$$L = \frac{1}{2}||\mathrm{w}||^2 - \sum_{\mathrm{i}=1}^{\mathrm{n}} \lambda i (y_i((w.x_i) + b) - 1 \tag{7.3}$$

where λi is the Lagrangian multiplier. The solution can be achieved by dual minimization of Eq. (7.3), with respect to w and b through standard procedures. For the non-separable case the constraints can be modified by introducing slack variables ζ_{i}.

$$y_i((w.x_i) + b) \geq 1 - \zeta_{\mathrm{i}} \tag{7.4}$$

and Eq. (7.1) becomes:

$$L = \frac{1}{2}||\mathrm{w}||^2 - \frac{1}{\mathrm{vn}} \sum_{\mathrm{i}=1}^{\mathrm{n}} \zeta_i \tag{7.5}$$

where v (0, 1) is introduced in order to account for misclassification (Scholkopf et al. 2000; Hastie et al. 2001). In addition, Vapnik (1995) introduced a kernel function K (x_i, x_j) to account for the non-linear decision boundary.

In this study, the Environment for Visualizing Images (ENVI 4.3) was used which provides four types of kernels functions i.e. linear, polynomial, radial basis function and sigmoid. Following steps should be taken to perform SVM model.

1. Preparation of the 15 landslide causative factors data layers.
2. Imported these factors in ENVI 4.3 software.
3. Defining region of interest (ROI) using landslide data.
4. Performing SVM classification algorithm for each factor by using the RBF kernel.
5. Summarizing the result of each factor.
6. Validation using area under curve (AUC) method.

7.2.2 Artificial Neural Network Model (ANNM)

Artificial Neural Network (ANN) is regarded as a machine which is framed in such a way that the human mind works to a specific task (Haykin 1999). The structure of ANN model is, the neurons which process the stimulus. Second one, each neuron has the ability to receive and transmit a signal which is called the stimulus. Next one, each signal—stimulus received or transmitted from one neuron to another in the neural network along with a weight (synaptic weight) which indicates the connective strength of the respective neurons (Mandal and Mondal 2019). The weight value is positively related to the contribution of the node (Fausett 1994). Even if the forms of relationship of the set of input factors are unknown or their physical explanation is hard to be described, ANN model has the ability to predict based on the input data. This characteristic of ANN model makes more superior than the empirical and statistical methods which need the prior knowledge of the input data and the nature of the relationship that is linear, non-linear etc. (Mandal and Mondal 2019).

7.2.2.1 The Learning Process Method

Through the iterative learning process, the neural networks receive stimuli (information and knowledge), as people do, and knowledge is stored in the network connections (Haykin 1999). The ANN models follow the parallel distributed processing which tries to combine the thinking of the human brain with the abstract mathematical thinking (McClelland et al. 1986). The learning method is grouped into three. Firstly, *the supervised learning* which is accomplished by presenting a set of training patterns associated with target output vector. Secondly, *the unsupervised learning* where learning is accomplished by grouping a similar set of input patterns together without the use of training data to specify what a typical member of each group looks like or to which group each pattern belongs to (Fausett 1994). Thirdly, *the reinforcement learning* is learning accomplished by interacting with an environment. During the training phase, the hidden and output layer neurons process their inputs by multiplying each input by a corresponding weight, summing the product, and then processing the sum using a non-linear transfer function to produce a result (Mandal and Mondal 2019).

7.2.2.2 The Basic Features of ANN Models

According to Fausett (1994) and Benardos and Benardos (2005), neural network comprises three basic features i.e. *network architecture*, *training* or *learning algorithm* and *activation function*. Network architecture is the way where all neurons are connected to each other. It is a critical process which requires possible knowledge of the problem application (Benardos and Kaliampakos 2004) while training or learning algorithm is the method which establishes the values of the weights on the connections. Last one is the type of activation function which is used to accomplish the model. Neural networks can be of single layer which resembles like a feed forward networks where the signals flow from the input units to the output units in a forward direction. In multilayer perceptron (MLP) networks there is always an *input layer*, a *hidden layer* and an *output layer* (Fig. 7.1). Input layer contains a node for each of the input variables, the hidden layer consists of nodes that allow complexities to develop among input nodes and output layer contains *t* nodes, one for each output type.

7.2.2.3 The Back-Propagation Algorithm

The training of an ANN by back-propagation learning algorithm dealt with three stages i.e. *feed forward of the input training pattern*, *calculation and back-*

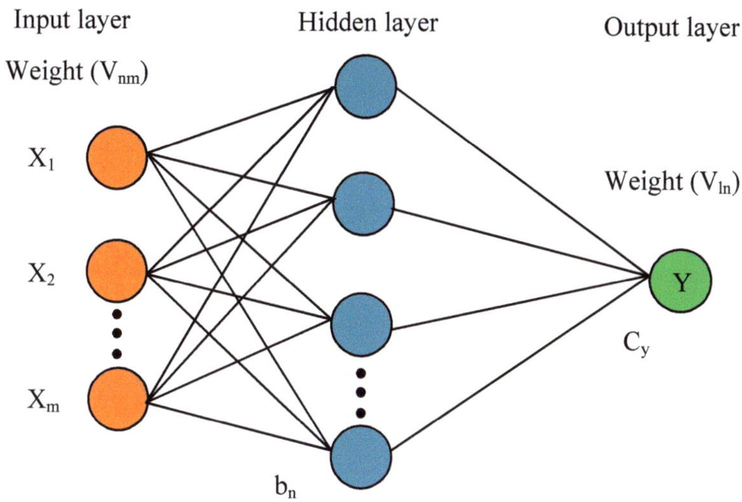

Fig. 7.1 The feed forward multi-layer artificial neural network

propagation of the associated error and *adjustment of the weights*. In first stage, each input unit receives an input signal and transmits this signal to each of the hidden units. Each hidden unit then calculates its activation and transmits its signal to each output unit (Eq. 7.6). Each output unit computes its activation to form the response of the net for the given input pattern (Eq. 7.7) (Mandal and Mondal 2019). Each neuron in the network may employ a nonlinear activation function at the output end for producing smooth signals to other neurons. One of most commonly used activation functions is the binary sigmoid transfer function which has range of (0, 1) and is defined by as Hagan et al. (1996) and Mandal and Mondal (2019) (Eq. 7.8). Each output unit compares it's activation with its target value to determine the associated error (Eq. 7.9). Based on this error, a δ factor, is being computed which is used to distribute the error in output unit back to all units in the hidden and input layer (Eqs. 7.10 and 7.11). After the entire δ factors have been calculated, the weights for all layers are adjusted simultaneously, according to the generalized Least-Square-Mean rue (Hagan et al. 1996) (Eq. 7.12).

$$net_j^{(l)}(t) = \sum_{i=0}^{p} (y_i^{(l-1)}(t) w_{ji}^{l}(t)), the\, net\, input\, of\; j{-}th\, neuron\; of\, layer\, l\, and\, t\, iteration \tag{7.6}$$

$$y_j^l(t) = f\left(net_j^{(l)}(t)\right) \tag{7.7}$$

$$f(net) = \frac{1}{1 + e^{(-net)}} \tag{7.8}$$

$$e_j(t) = c_j(t) - a_j(t) \tag{7.9}$$

$$\delta_j^{(l)}(t) = e_j^{(l)}(t)a_j(t)\left[1 - a_j x(t)\right], \\ \delta\, factor\ for\, neuron\ j{-}th\, in\, the\, output\ layer\ i{-}th \tag{7.10}$$

$$\delta_j^{(l)}(t) = y_j^{(l)}(t)\left[1 - y_j(t)\right]\sum \delta_j^{(l)}(t)w_{kj}^{(l+1)}(t), \\ \delta\, factor\ for\, neuron\ j{-}th\, in\, hidden\, layer\ i{-}th \tag{7.11}$$

$$w_{ji}^{(l)}(t+1) = w_{ji}^{(l)}(t) + \alpha\left[w_{ji}^{(l)}(t) - w_{ji}^{(l)}(t-1)\right] + n\delta_j^{(l)}(t)y_j^{(l-1)}(t) \tag{7.12}$$

where, η is the learning rate, and α is the momentum rate for speeding up learning without running into the risk of oscillation.

7.2.2.4 The Process of Landslide Susceptibility Analysis Through ANN Models in MATLAB Environment

The probabilities of occurrence of landslides were estimated based on the various attributes and their cumulative influence. Before running the artificial neural network model, the training site was selected. So, the landslide affected (occurrence) area and the landslide non-affected area were selected as training sites. There was no precise mathematical law to establish the required minimum size of landslide affected datasets. In this study, about 70% of the dataset was considered as sufficient for training the network and the rest of it is normally considered for testing the model. Pixels from each of the two classes were randomly selected as training pixels denoting areas where landslides occurred or not occurred. Firstly, areas where landslide was not occurred were classified as "areas not prone to landslide" and areas where landslide activities existing assigned as an "areas prone to landslide" training set. Training sites were considered based on landslide locations as prone training site and with a varying slope values as non-prone training site and then the MLP trained back propagation algorithm was computed. Hidden layers were selected two times of input attribute layers. The output will have both "existing" and "non-existing" landslide areas. Some of the input attributes layers are continuous and others categorical in nature (Mandal and Mondal 2019). Therefore, these data were converted to raster grid in order to apply the ANN model. The model had 15 neurons in the input layer, with 30 hidden neurons. The 15 neurons in the input layer correspond to the landslide related factors (elevation, slope, aspect, curvature, lineament, distance to lineament, drainage density, distance to drainage, NDVI, land use and land cover, geology, soil, rainfall, SPI and TWI) that had been identified as causative factors. Three-layered feed-forward network was implemented using the MATLAB software package. The number of hidden layers and the number of nodes in a hidden layer required for a particular classification problem are not easy to deduce. In this study, an input layer (15 input neurons) × one hidden layer (30 hidden neurons) × one output layer (2 output neurons) is selected for the network model of 15-30-2, with input data normalized in the range 0.1–0.9. The nominal and interval class group data were converted to continuous values ranging between 0.1 and 0.9. Therefore, all the layers were normalized in the range 0.1–0.9. The categorical data and their interval class group were converted to a continuous values ranging 0.1–0.9. In this way, the continuous values became nominal for back propagation modelling. The learning rate was set to 0.01, and the initial weights were randomly selected between 0.1 and 0.9. The MLP trained back-propagation algorithm was used to minimize the error between the predicted output values and the calculated output values. The algorithm propagated the error backwards iteratively by adjusting the weights. The number of epochs was set up to 10,000 and the root mean square error (RMSE) value used for the stopping criterion was set to 0.01. Most of the training datasets met the 0.011 RMSE goal. However, the maximum number of iterations was set to 10,000 epochs. When the latter case occurred, then the maximum RMSE value was 0.173 which selected for this study. Finally, the landslide susceptibility maps were generated and classified as very high, high, moderate, low and very low susceptibility classes (Mandal and Mondal 2019).

7.3 Support Vector Machine, Landslide Susceptibility Zones and Frequency Ratio (FR)

The Balason river basin was classified into five landslide susceptibility zones i.e. very low, low, moderate, high and very high landslide susceptibility which is based on SVM model (Fig. 7.2a). Out of total area of the Balason river basin, 24.97% area of the whole basin is under moderate landslide susceptibility, 21.97% is under high susceptibility, 10.58% under high landslide susceptibility and 17.97% under low landslide susceptibility. On the other hand, the study also revealed that 53.07% area is affected by landslides in very high landslide susceptibility zone and 27.17% area is affected by landslides in high landslide susceptibility zones (Table 7.1). The frequency ratio is very high (5.02) in very high landslide susceptibility zone which also depicts high chances of landslides.

Table 7.1 SVM, landslide susceptibility and frequency ratio

Landslide susceptibility zones	Total pixels	% of total pixels	Landslide pixels	% of landslide pixels	Frequency ratio
Very low	75,571	17.97	0	0.00	0.00
Low	103,049	24.51	22	1.63	0.07
Moderate	105,008	24.97	248	18.13	0.73
High	92,373	21.97	372	27.17	1.24
Very high	44,499	10.58	726	53.07	5.02

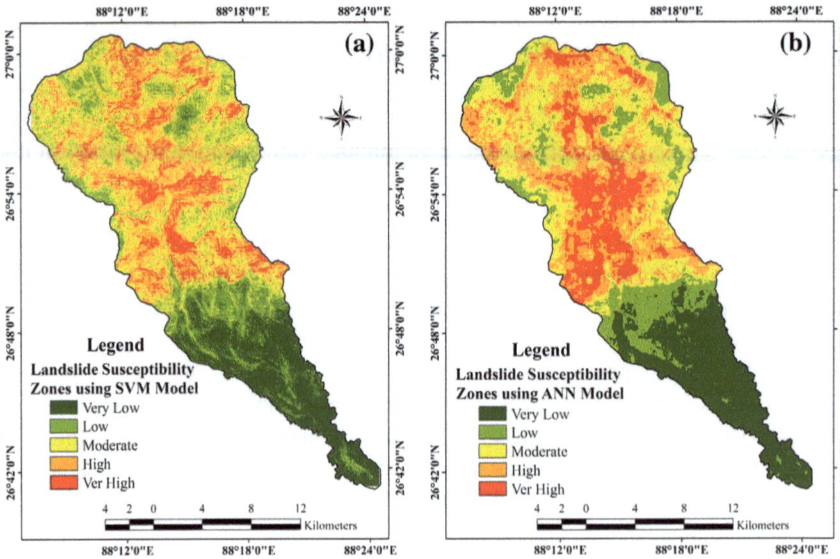

Fig. 7.2 Landslide susceptibility zones of the Balason river basin. **a** SVM model and **b** ANN model

7.4 Artificial Neural Network Model (ANNM), Landslide Susceptibility Zones and Frequency Ratio (FR)

Artificial Neural Network Model base landslide susceptibility zonation map divided the Balason river basin into five distinct zones i.e. very low, low, moderate, high and very high landslide susceptibility (Fig. 7.2b). According to ANN model, the Balason river basin is dominated by high landslide susceptibility which is followed by moderate, very low, low and very high landslide susceptibility (Fig. 7.3). The study demonstrated that 43.68% area is affected by landslides in very high landslide susceptibility zones, 31.39% area is affected by landslides in high landslide susceptibility zone and only around 7% area is affected by landslides in low and very low landslide susceptibility zones (Table 7.2). In the Balason river basin, very high landslide susceptibility zone is attributed with high frequency ratio value of 2.64 and low landslide susceptibility zone is registered with low frequency ratio (0.01) and low landslide probability.

7.5 Validation Results of SVM and ANN Model

An accuracy study has been made to assess the validation of landslide susceptibility maps. In this study ROC has been used for measuring the accuracy level of the landslide susceptibility zonation map. The ROC curve is the graphical presentation of the sensitivity on the y axis and 1-specificity of the model on the

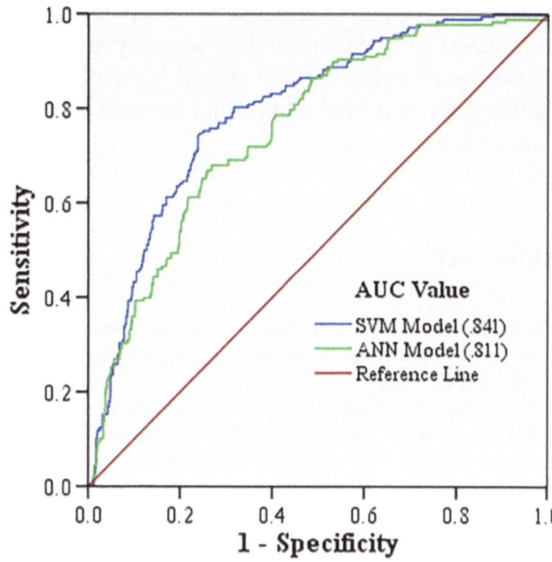

Fig. 7.3 ROC curve of SVM and ANN model based landslide susceptibility map

Table 7.2 ANN Model, Landslide susceptibility and frequency ratio (FR)

Landslide susceptibility zones	Total pixels	% of total pixels	Landslide pixels	% of landslide pixels	Frequency ratio
Very Low	84,305	20.05	2	0.14	0.01
Low	70,212	16.70	91	6.66	0.40
Moderate	92,923	22.10	248	18.14	0.82
High	103,426	24.60	429	31.39	1.28
Very High	69,634	16.56	598	43.68	2.64

x axis (Mohammady et al. 2012). The area under curve for SVM based landslide susceptibility zonation map is 0.841 (84.1%) and for ANN model based landslide susceptibility map is 0.811 (81.1%). Between SVM and ANN model, SVM model depicted high level of accuracy and this model can be used for the preparation of landslide susceptibility zonation map.

7.6 Conclusion

The main objective of the study was to perform SVM model and ANN model using geoinformatics and to prepare landslide susceptibility zonation maps of the Balason river basin of Darjeeling Himalaya. The study identified potential landslide susceptibility locations analysing various geomorphic, geohydrologic, and tectonic parameters. The prepared landslide susceptibility zonation map of the Balason river basin based on SVM shown 84% accuracy and which could be accepted and used for site-specific planning and management of hilly terrain of the basin. Hydrology and lithology of the basin plays an important role to make the mountain slope more vulnerable to slope failure.

References

Ada M, San BT (2018) Comparison of machine-learning techniques for landslide susceptibility mapping using two-level random sampling (2LRS) in Alakir Catchment area, Antalya, Turkey. Nat Hazards 90:237–263

Benardos AG, Kaliampakos DC (2004) A methodology for assessing geotechnical hazards for TBM tunnelling—illustrated by Athens Metro, Greece. Int J Rock Mech Min Sci 41(4):987–999

Benardos AG, Benardos AP (2005) Applications of artificial neural networks in geotechnology. Chronicles of Mining and Metallurgical 15(1):65–81 (in Greek)

Bui DT et al (2012) Spatial prediction of landslide hazards in HoaBinh province (Vietnam): a comparative assessment of the efficacy of evidential belief functions and fuzzy logic models. Catena 96:28–40

Caparrini F, Caporali E, Castelli F (1996) Neural network analysis of satellite images for land cover discrimination. In: Claps P, Siccardi F (eds) Mediterranean storms. Proceedings of the EGS Plinius conference held at Maratea, Italy, Oct 1999, pp 603–613

Carrara A, Cardinali M, Detti R, Guzzetti F, Pasqui V, Reichenbach P (1991) GIS techniques and statistical models in evaluating landslide hazard. Earth Surf Proc Land 16:427–445

Carrara A, Cardinali M, Guazzetti F, Reichenbach P (1995) GIS techniques in mapping landslide hazard. In: Carrara A, Guzzetti F (eds) Geographical information systems in assessing natural hazards. Kluwer Academic Publishers, The Netherlands, pp 135–175

Emami SMR, Iwao Y, Harada T (1998) Performance of Artificial Network system in prediction issues of earthquake engineering. In: Proceedings of 8th international IAEG congress, pp 733–738

Fausett L (1994) Fundamentals of neural networks architectures. Algorithms and applications, Prentice Hall, USA, p 461

Goetz J, Brenning A, Petschko H, Leopold P (2015) Evaluating machine learning and statistical prediction techniques for landslide susceptibility modeling. Comput Geosci 81:1–11

Hagan TM, Demuth BH, Beale HM (1996) Neural network design. In: Electrical Engineering Series, Brooks/Cole, p 730

Hastie T, Tibshirani R, Friedman J (2001) The elements of statistical learning: data mining inference and prediction. Springer, New York, NY, USA

Haykin S (1999) Neural networks a comprehensive foundation, 2nd edn. Prentice Hall, Englewood Cliffs, New Jersey, USA, p 696

Hong et al (2005) The influence of intense rainfall on the activity of large scale crystalline schist landslides in Shikoku Island, Japan. Landslides 2:97–105

Ilanloo M (2011) A comparative study of fuzzy logic approach for landslide susceptibility mapping using GIS: an experience of Karaj dam basin in Iran. Proc Soc Behav Sci 19:668–676

Lee S, Ryu J, Min K, Won J (2001) Proceedings of the geoscience and remote sensing symposium, IGARSS '01, IEEE 2001 international, vol 5, pp 2364–2366

Lee S, Ryu JH, Lee MJ, Won JS (2003) Landslide susceptibility analysis using artificial neural network at Boeun, Korea. Environ Geol 44:820–833

Lee S, Ryu JH, Won JS, Park HJ (2004) Determination and application of the weights for landslide susceptibility mapping using an artificial neural network. Engr. Geol. 71:289–302

Lee S (2007) Application and verification of fuzzy algebraic operators to landslide susceptibility mapping. Environ Geol 52:615–623

Mayoraz F, Cornu T, Vuillet L (1996) Using neural networks to predict slope movements. In: Proceedings of VII international symposium on landslides, Trondheim, June 1966, 1. Balkema, Rotterdam, pp 295–300

Mandal, Mondal (2019) Statistical approaches for landslide susceptibility assessment and prediction,. Springer International Publishing AG, Part of Springer Nature. https://doi.org/10.1007/978-3-319-93897-4

McClelland JL, Rumelhart DE, the PDP research group, (1986) Parallel distributed processing: explorations in the microstructure of cognition, vol II. MIT Press, Cambridge, MA

Mohammady M, Pourghasemi HR, Pradhan B (2012) Landslide susceptibility mapping at Golestan Province, Iran: A comparison between frequency ratio, Dempster-Shafer, and weights-of-evidence models. J Asian Earth Sci 61:221–236. https://doi.org/10.1016/j.jseaes.2012.10.005

Pham BT, Tien Bui D, Pham HV (2016) Spatial prediction of rainfall induced landslides using Bayesian Network at Luc Yen District, Yen Bai Province (Viet Nam). In: International conference on environmental issues in mining and natural resources development (EMNR 2016). Hanoi University of mining and geology (HUMG), Vietnam, pp 1–10

Pistocchi A, Luzi L, Napolitano P (2002) The use of predictive modelling techniques for optimal exploitation of spatial databases: a case study in landslide hazard mapping with expert system-like methods. Environ Geol 41:765–775

Pradhan B (2010a) Remote sensing and GIS-based landslide hazard analysis and cross-validation using multivariate logistic regression model on three test areas in Malaysia. Adv Space Res 45:1244–1256

Pradhan B (2010b) Use of GIS-based fuzzy logic relations and its cross application to produce landslide susceptibility maps in three test areas in Malaysia. Environ Earth Sci (article on-line first available). https://doi.org/10.1007/s12665-010-0705-1

Pradhan B (2010c) Manifestation of an advanced fuzzy logic model coupled with geo-information techniques to landslide susceptibility mapping and their comparison with logistic regression modelling. Environ Ecol Stat (article on-line first available). https://doi.org/10.1007/s10651-010-0147-7

Pradhan BA (2013) Comparative study on the predictive ability of the decision tree, support vector machine and neuro-fuzzy models in landslide susceptibility mapping using GIS. Comput Geosci 51:350–365

Scholkopf B, Smola A, Williamson RC, Bartlett PL (2000) New support vector algorithms. Neural Comput 12:1207–1245

Sezer EA, Pradhan B, Gokceoglu C (2011) Manifestation of an adaptive neuro-fuzzy model on landslide susceptibility mapping: Klang valley, Malaysia. Expert Syst Appl 38(7):8208–8219

Tsangaratos P, Ilia I (2016) Landslide susceptibility mapping using a modified decision tree classifier in the Xanthi Perfection, Greece. Landslides 13(2):305–320

Vapnik V (1995) The nature of statistical learning theory. Springer, New York, NY, USA

Wu X, Ren F, Niu R (2014) Landslide susceptibility assessment using object mapping units, decision tree, and support vector machine models in the Three Gorges of China. Environ Earth Sci 71(11):4725–4738

Xie M, Esaki T, Cai M (2004) A time-space based approach for mapping rainfall-induced shallow landslide hazard. Earth Surf Proc Land 26:1251–1263

Xu C, Xu X, Dai F, Saraf AK (2012a) Comparison of different models for susceptibility mapping of earthquake triggered landslides related with the 2008 Wenchuan earthquake in China. Comput Geosci 46:317–329

Xu C, Dai F, Xu X, Lee YH (2012b) GIS-based support vector machine modeling of earthquake-triggered landslide susceptibility in the Jianjiang River watershed, China. Geomorphology 145–146:70–80

Yao X, Tham LG, Dai FC (2008) Landslide susceptibility mapping based on Support Vector Machine: a case study on natural slopes of Hong Kong, China. Geomorphology 101:572–582

Zhu A-X et al (2014) An expert knowledge-based approach to landslide susceptibility mapping using GIS and fuzzy logic. Geomorphology 214:128–138

Chapter 8
Factor's Clustering and Identification of Suitable Factor's Group Model in Landslide Susceptibility

8.1 Introduction

The occurrences of landslides over the hilly landscape of the Darjeeling Himalaya are the most serious concern which creates constant problems and causing serious damages to properties and lives as well as hampering the communication line in every year. Balason river basin is not an exception of these phenomena. Several literatures on landslides of Darjeeling Himalaya (Nautiiyal 1966; Basu 1970; Basu and Sarkar 1987; Sengupta 1995; Kanungo et al. 2006; Mandal and Maiti 2014) were already stated about the enormity of landslides damages done. However, the process has been accelerated by extensive deforestation, unsystematic and unscientific constructional activities, inadequate drainage system, urban expansion, rapid changes of land use practice etc. In this region, landslides commonly occurred on cut slopes, embankments alongside roads and highways and also near the high-rise apartments and residential areas (Cheng et al. 2008). Various causative variables, which affect slope instability and cause landslides, are geology, hydrology, hydrogeology, topography and morphology, climate and weathering (Soeters and Van Westen 1996; Chau et al. 2004; Peart et al. 2005; Dominguez-Cuesta et al. 2007; Garfi and Bruno 2007). Three important topographic factors like elevation, slope and directionality of slope influence on landslide occurrence (Dai and Lee 2002; Chau and Chan 2005). Researchers like Can et al. (2005), Lee and Sambath (2006) and Greco et al. (2007) considered slope curvature as an important topographic attribute for landslides. On the other hand, slope instability occurs due to heavy or long duration rainfall, water infiltration into cracks (Crosta 1998; Corominas and Moya 1999; Ayalew et al. 2005; Dahal and Hasegawa 2008), presence of large active faults, streams activities (Saha et al. 2002; Yaclin 2008; Kelarestaghi and Ahmadi 2009), road construction in mountainous lands (Greco et al. 2007; Lee and Pradhan 2007), land cover and land use changes (Dai et al. 2004; Lee and Sambath 2006; Greco et al. 2007), weak lithological and structural characteristics of the region concerned (Garcia-Rodriguez et al. 2008;

S. Mandal and S. Mondal, *Geoinformatics and Modelling of Landslide Susceptibility and Risk*, Environmental Science and Engineering,
https://doi.org/10.1007/978-3-030-10495-5_8

Yaclin 2008). Many researcher prepared landslide susceptibility map by considering these factors for the better management of soil, water and natural resources. In this study spatial relationship was assessed to show the relative importance of each factor group model (geomorphological factors group, lithological factors group, hydrological factors group, triggering factor, protective factor and anthropogenic factor models) in landslide susceptibility assessments based on the certainty factor (CF) model. CF model has been because of its high accuracy in landslide susceptibility assessment and prediction.

8.2 Materials and Methods

The present study includes the preparation of various data layers and their groupings considering geomorphological factors, lithological factors, hydrological factors, triggering factors and anthropogenic factors. The integration between various data layers of certain group was made applying weighted linear combination model and six separate models were being performed i.e. geomorphological factors group (GFG) model, lithological factors group (LFG) model, hydrological factors group (HFG) model, triggering factor (TF) model, protective factor (PF) model and anthropogenic factor (AF) model. The main objective to prepare these models was to establish the relationship of each model and integrated layer with landslide susceptibility (LS) index map using *Certainty Factor Model* (CFM) and its respective parameters. The statistical correlation analysis was performed in Arc GIS 10.1 platform.

8.2.1 Factor's Group Model and Associated Data Layers

In this study, an attempt has been made to prepare six different weighted linear factors group models by considering the homogeneity within landslide causative factors (CFs). These factors group models were geomorphological factors group (GFG) model, lithological factors group (LFG) model, hydrological factors group (HFG) model, triggering factor (TF) model, protective factor (PF) model and anthropogenic factor (AF) model (Figs. 8.1, 8.2 and 8.3). The GFG model was calibrated using elevation (El), aspect (As), slope (Sl) and curvature (Cu) layers. The LFG model was prepared considering geology (Ge), soil (So), lineament density (Ld) and distance to lineament (D2l) factors. For HFG model drainage density (Dd), distance to drainage (D2d), stream power index (SPI) and topographic wetted index (TWI) data layers were taken into account. TF, PF and AF models were generated by considering rainfall (Ra), normalized differential vegetation index (NDVI) and (land use and land cover) LULC data layer respectively.

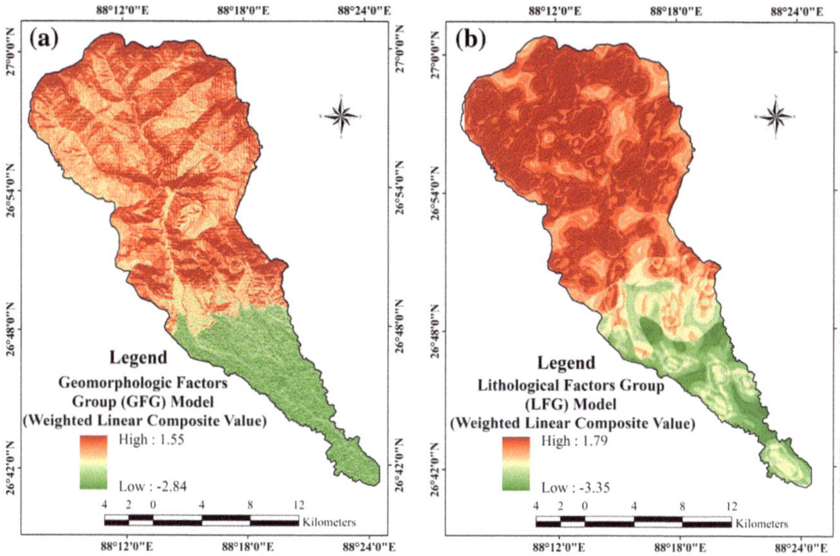

Fig. 8.1 Factors group models **a** GFG model and **b** LFG model

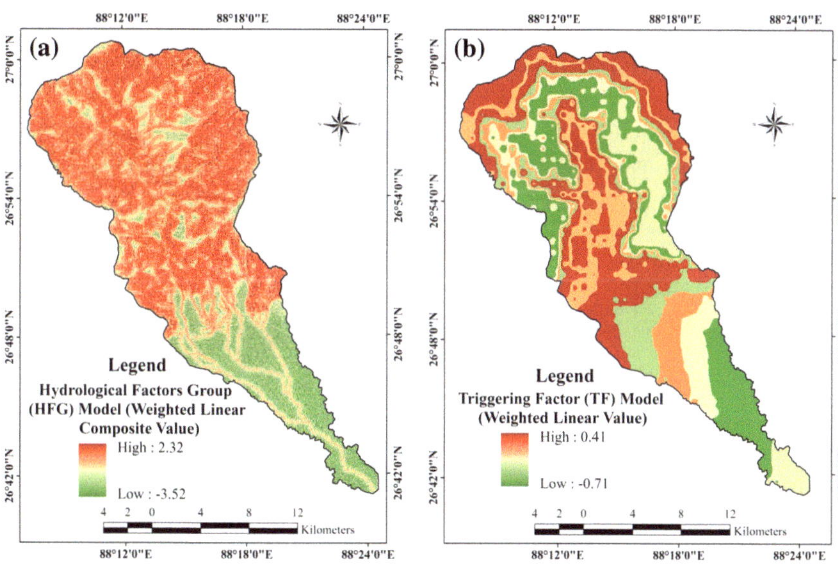

Fig. 8.2 Factors group models **a** HFG model and **b** TF model

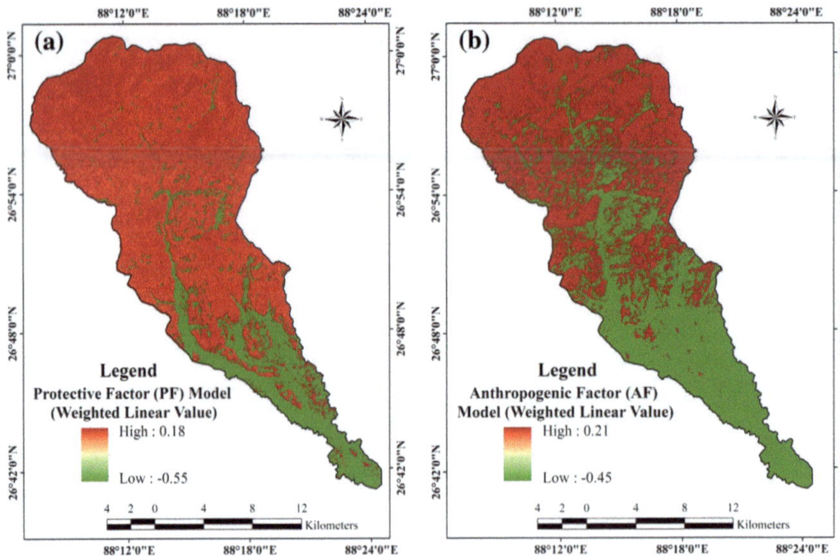

Fig. 8.3 Factors models **a** PF model and **b** AF model

8.2.2 Certainty Factor Model (CFM)

Certainty factor approach is dealt with the problem of combination of non-homogeneous data. This model has been successfully used by Kanungo et al. (2011), Sujatha et al. (2012) and Liu et al. (2014) in landslide susceptibility mapping studies.

$$\mathrm{CF} = \begin{cases} \dfrac{ppa - pps}{ppa(1 - pps)} & if\ ppa \geq pps \\ \dfrac{ppa - pps}{ppa(1 - ppa)} & if\ ppa < pps \end{cases} \tag{8.1}$$

where CF is the certainty factor (Chap. 6, Table 6.1), ppa is the conditional probability of having a number of landslides in a class 'a' and pps is the prior probability of having the total number of landslides in the study area 'A'.

The layers are combined pairwise according to the integration rules (Chung and Fabbri 1993; Binaghi et al. 1998). The combination of CF values of two thematic layers 'z' is expressed in the following equation as given by Binaghi et al. (1998).

$$\mathrm{Z} = \begin{cases} x + y - xy,\ x, y \geq 0 \\ \frac{x+y}{1-\min(|x|\cdot|y|)}\ x, y \text{ opposite sign} \\ x + y + xy,\ x, y < 0 \end{cases} \tag{8.2}$$

8.2.3 *Landslide Susceptibility Zones and Frequency Ratio*

The CFM induced landslide susceptibility zonation map of the Balason river basin is dominated by high landslide susceptibility (33.93%) which is followed by very high, moderate, low and very low landslide susceptibility (Chap. 6, Fig. 6.1b). Very high and high landslide susceptibility zones are highly affected by landslides where 50% area is affected by landslides in very high susceptibility zones (Chap. 6, Table 6.3). On the other hand low and very low susceptibility zones are least affected by landslide phenomena. It can be assumed that CFM has provided a result which can be accepted a best suited landslide model of Darjeeling Himalaya. The moderate, high and very high landslide susceptibility zones are registered with frequency ratio value of 0.34, 1.30 and 1.88 respectively.

The validation of the landslide susceptibility zonation map based certainty factor model was accomplished using ROC curve called area under curve value (AUCV) which represents the predictive capability of the landslide susceptibility map. The certainty factor model showed 0.849 or 84% for (CFM). The result of ROC curve depicted that certainty factor model (CFM) has high % accuracy (84%) and may be accepted as one of the most significant bivariate model in landslide susceptibility studies (Chap. 6, Sect. 6.4).

8.2.4 *Statistical Correlation Between Landslide Susceptibility Map (CFM) and GFG Model, LFG Model, HFG Model, TF Model, PF Model and AF Model*

To assess or identify the suitable factor group model a statistical correlation analysis was made between *certainty factor model* based landslide susceptibility map and individual factor group model (GFG model, LFG model, HFG model, TF model and AF model) on GIS environment. This correlation was performed in ARC GIS 10.1 using PCA tool.

8.3 Results and Discussion

It is already stated that, 6 different factors group models were used to determine the role of individual factor group model in assessing LS. LFG model revealed most intimate association between landslide susceptibility and LFG parameters having correlation value of 0.86 (Table 8.2). In this model, Ge factor (0.81) played most dominant role to determine vulnerability of landslides followed by So (0.80), Ld (0.29) and D2l (0.24). Strong relationship with LS was observed in case of GFG model (0.85) followed by hydrological factor group model (0.83). Among the geomorphologic factors, El and Sl has the greater correlation value of 0.83 followed

Table 8.1 Spatial correlation between landslide susceptibility index model and GFG model

Model	GFG model				
Landslide susceptibility index (LSI) model	El	As	Sl	Cu	Model
	0.83	0.11	0.83	−0.02	0.85

Table 8.2 Spatial correlation between landslide susceptibility index model and LFG model

Model	LFG model				
Landslide susceptibility index (LSI) model	Ge	So	Ld	D2l	Model
	0.81	0.80	0.29	0.24	0.86

Table 8.3 Spatial correlation between landslide susceptibility index model and HFG model

Model	HFG model				
Landslide susceptibility index (LSI) model	Dd	D2d	SPI	TWI	Model
	0.65	0.56	0.69	0.55	0.83

Table 8.4 Spatial correlation between landslide susceptibility index model and TF model, PF model and AF model

Model	TF model	PF model	AF model
Landslide susceptibility index (LSI) model	Ra/Model	NDVI/Model	LULC/Model
	0.40	0.66	0.67

by As (0.11) and Cu (−0.02) (Table 8.1). In hydrological factor group model, SPI (0.69) has greater significance to assess LS followed by Dd (0.65), D2d (0.56), and SPI (0.55) (Table 8.3). PF and AF models have the correlation value of 0.66 and 0.67 respectively. Lastly, TF model exerted lowest influence on LS as indicated by correlation value of 0.40 (Table 8.4).

The factors associated with respective factors group model do not have similar impact to prepare factors group models. El has the highest relationship (0.84) with GFG model followed by Sl (0.83), As (0.42) and Cu (−0.14). Similarly, Ge (0.82) and Dd (0.77) have greater relationship with LFG model and hydrological factor group model respectively (Table 8.5).

8.4 Conclusion

It can be concluded that each model can be treated as unique in terms of terrain processes and forms and can be used to identify role of geomorphological factors group model, lithological factors group model, hydrological factors group model, triggering factor model, protective factor model and anthropogenic factor model in slope failures individually. Lithological factors group model was played most

Table 8.5 Spatial correlation between factors group models and the parameters associated with the respective models

Model	El	As	Sl	Cu	Ge	So	Ld	D2l	Dd	D2d	SPI	TWI
GFG model	0.84	0.42	0.83	−0.14	–	–	–	–	–	–	–	–
LFG model	–	–	–	–	0.82	0.78	0.55	0.50	–	–	–	–
HFG model	–	–	–	–	–	–	–	–	0.77	0.75	0.70	0.60

prominent role to determine the landslides vulnerability zones as indicated by correlation value of 0.86 followed by geomorphological factors group model (0.85), hydrological factor group model (0.83), anthropogenic factor model (0.67), protective factor model (0.66) and triggering factor model (0.40).

References

Ayalew L, Yamagishi H, Marui H, Kanno T (2005) Landslides in Sado Island of Japan: part II. GIS-based susceptibility mapping with comparisons of results from two methods and verifications. Eng Geol 81:432–445

Basu AR (1970) A geotechnical note on investigations of certain stretches of Border roads under Swastik Project in Sikkim and Darjeeling District, West Bengal. Unpublished GSI report, FS 1969–1970

Basu SR, Sarkar S (1987) Ecosystem vis-a-vis landslide: a case study in Darjeeling Himalaya, Proce. Impact of development on environment. Geog Soc India 2:45–53

Binaghi E, Luzi L, Madella P (1998) Slope instability zonation: a comparison between certainty factor and fuzzy Dempster–Shafer approaches. Nat Hazards 17:77–97

Can T, Nefeslioglu HA, Gokceoglu C, Sonmez H, Duman TY (2005) Susceptibility assessments of shallow earth flows triggered by heavy rainfall at three catchments by logistic regression analyses. Geomorphology 72:250–271

Chau KT, Chan JE (2005) Regional bias of landslide data in generating susceptibility maps using logistic regression for Hong Kong Island. Landslides 2:280–290

Chau KT, Sze YL, Fung MK, Wong WY, Fong EL, Chan LCP (2004) Landslide hazard analysis for Hong Kong using landslide inventory and GIS. Comput Geosci 30:429–443

Cheng TA, Lateh H, Peng KS (2008) Intelligence explanation system on landslide dissemination: a case study in Malaysia. In: Proceedings of the first world landslide forum report: implementing the 2006 Tokyo action plan on the international program on landslides (IPL), pp 330–333

Chung CF, Fabbri AG (1993) The representation of geosciences information for data integration. Nonrenew Resour 2(2):122–139

Corominas J, Moya J (1999) Reconstructing recent landslide activity in relation to rainfall in the Llobregat river basin, Eastern Pyrenees, Spain. Geomorphology 30:79–93

Crosta G (1998) Regionalization of rainfall thresholds: an aid to landslide hazard evaluation. Environ Geol 35:131–145

Dahal RK, Hasegawa SH (2008) Representative rainfall thresholds for landslides in the Nepal Himalaya. Geomorphology 100:429–443

Dai FC, Lee CF (2002) Landslide characteristics and slope instability modeling using GIS, Lantau Island, Hong Kong. Geomorphology 42:213–228

Dai FC, Lee CF, Tham LG, Ng KC, Shum WL (2004) Logistic regression modelling of storm-induced shallow landsliding in time and space on natural terrain of Lantau Island, Hong Kong. Bull Eng Geol Environ 63:315–327

Dominguez-Cuesta M, Jimenez-Sanchez M, Berrezueta E (2007) Landslides in the central coal-field (Cantabrian Mountains, NW Spain): geomorphological features, conditioning factors and methodological implications in susceptibility assessment. Geomorphology 89:1–12

Garcia-Rodriguez MJ, Malpica JA, Benito B, Diaz M (2008) Susceptibility assessment of earthquake-triggered landslides in El Salvador using logistic regression. Geomorphology 95:172–191

Garfi G, Bruno DE (2007) Fan morphodynamics and slope instability in the Mucone River basin (Sila Massif, southern Italy): significance of weathering and role of land use changes. Catena 50:181–196

Greco R, Sorriso-Valvo M, Catalano E (2007) Logistic regression analysis in the evaluation of mass movements susceptibility: the Aspromonte case study, Calabria, Italy. Eng Geol 89:47–66

Kanungo DP, Arora MK, Sarkar S, Gupta RP (2006) A comparative study of conventional, ANN black box, fuzzy and combined neural and fuzzy weighting procedures for landslide susceptibility zonation in Darjeeling Himalayas. Eng Geol 85(3–4):347–366

Kanungo DP, Sarkar S, Sharma S (2011) Combining Neural Network with fuzzy, certainty factor and likelihood ratio concepts for spatial prediction of landslide. Nat Hazards 59(3):1491–1512

Kelarestaghi A, Ahmadi H (2009) Landslide susceptibility analysis with a bivariate approach and GIS in northern Iran. Arab J Geosci 2:95–101

Lee S, Pradhan B (2007) Landslide hazard mapping at Selangor, Malaysia using frequency ratio and logistic regression models. Landslides 4:33–41

Lee S, Sambath T (2006) Landslide susceptibility mapping in the Damrei Romel area, Cambodia using frequency ratio and logistic regression models. Environ Geol 50:847–855

Liu M, Chen X, Yang S (2014) Collapse landslide and mudslide hazard zonation. In: Landslide science for a safer geoenvironment. Springer International Publishing, pp 457–462

Mandal S, Maiti R (2014) Role of lithological composition and lineaments in landsliding: a case study of Shivkhola watershed, Darjeeling Himalaya. Int J Geol Earth Environ Sci. 4(1):126–132

Nautiiyal SP (1966) On the stability of certain hill slopes in and around Darjeeling, W.B. Bull Geol Surv India 15:31–48

Peart MR, Ng KY, Zhang DD (2005) Landslides and sediment delivery to a drainage system: some observations from Hong Kong. Asian Earth Sci 25:821–836

Saha AK, Gupta RP, Arora MK (2002) GIS-based landslide hazard zonation in the Bhagirathi (Ganga) valley, Himalayas. Int J Remote Sens 23:357–369

Sengupta CK (1995) Detailed study of geofactors in selected hazard prone stretches along the surface communication routes in parts of Darjeeling and Sikkim Himalaya, Phase-I, Part-I (Rongtong-Kurseong road section). Unpublished GSI report, FS 1993–94 & 1994–95

Sujatha ER, Rajamanikam GV, Kumaravel P (2012) Landslide susceptibility analysis using probabilistic certainty factor approach: a case study on Tevankarai stream watershed, India. J Earth Syst Sci 121(5):1337 1350

Soeters R, Van Westen CJ (1996) Slope instability recognition, analysis and zonation. In: Turner AK, Schuster RL (eds) Landslides: investigation and mitigation. Transportation Research Board, National Research Council, Special Report 247. National Academy Press, Washington, DC, pp 129–177

Yaclin A (2008) GIS-based landslide susceptibility mapping using analytical hierarchy process and bivariate statistics in Ardesen (Turkey): comparisons of results and confirmations. Catena 72:1–12

Chapter 9
Landslide Susceptibility and Elements at Risk: A Brief Review

9.1 Introduction

Since ancient time, the Darjeeling Himalaya has been subjected to landslide hazards and similar phenomena. The Balason river basin is situated in this region and suffered several types of large and small landslides. At the time of pre-independence landslides took place mostly in the areas where the setup of the physical environment was weak, less strong and more disturbed geological formations, rugged and high slope areas, low vegetation cover areas etc. These natural geo-environmental parameters play an important role in slope instability (Hays 1981; Sarkar and Kanungo 2010). In recent times, the severity and the frequency of the landslide phenomena have been increased significantly. The major reason behind it unscientific and unsystematic settlement construction due to unprecedented growth of population, deforestation, overloaded vehicles on the roads which induced tremor, lack of proper maintenance of tea garden areas, dumping of garbage behind the built up areas (Bhattacharya 1998, 1999, 2002, 2010) along with the aforementioned factors. Landslides become a problem when they interfere with human lives and activities. Every year landslides cause serious problems to the local habitation and communication system especially in the rainy season. It is therefore essential to analyze the susceptibility of landslide occurrences for the Balason river basin. Many researchers carried out demand-based studies in different mountainous and hilly tracts of India regarding landslide causative factors, different landslide susceptible and risk areas and mitigation measures (Nautiyal 1951; Sinha 1974; Sengupta 1995; Bhandari 2007; Mandal and Maiti 2014; Mandal and Mandal 2017). However, there are no general technique exist for the preparation of landslide susceptibility maps (Carrara et al. 1995; van Westen 1997). The methods of landslide susceptibility mapping can be qualitative and quantitative in nature, for example, heuristic analysis (geomorphic analysis and qualitative map combination), knowledge based analysis (relating to learning process), statistical analysis (frequency ratio analysis, information value method, weights of evidence

S. Mandal and S. Mondal, *Geoinformatics and Modelling of Landslide Susceptibility and Risk*, Environmental Science and Engineering,
https://doi.org/10.1007/978-3-030-10495-5_9

method landslide normal risk factor, multiple logistic regression, discriminant analysis) and deterministic analysis (factor of safety approach and classical slope stability theory) (Fell et al. 2008). In Chap. 5, several bivariate models were used to prepare landslide susceptibility map i.e. modified information value (MIV) model, landslide nominal/normal/numerical risk factor (LNRF), fuzzy logic (FL) and statistical index (SI) Model. In Chaps. 7 and 8, multivariate models (Bayesian probability based weights of evidence model and certainty factor) and machine learning models (support vector machine and artificial neural network) were employed respectively. Among these, certainty factor model yielded very good landslide prediction accuracy of 84.90%. This study tried to focus on the assessment of the susceptibility of landslide occurrences in the light of human settlement area and metalled road based on the certainty factor (CF) model so that their bad effects on settlement area and road are amended right way.

9.2 Materials and Methods

Data acquisition and construction of base maps regarding the distribution of landslides and the influencing factors of landslides is very much important for the landslide susceptibility mapping (Ermini et al. 2005; Ercanoglu 2005). In the study, fifteen landslide inducing factors were considered i.e. elevation, aspect, slope, curvature, geology, soil, lineament density, distance to lineament, drainage density, distance to drainage, stream power index (SPI), topographic wetted index (TWI), rainfall (Ra), normalized differential vegetation index (NDVI) and land use and land cover (LULC) for the preparation landslide susceptibility zonation maps using CF model. The assessment of different susceptibility categories of settlement area and metalled road were done based on the aforesaid susceptibility zonation method. The elements at risk data layers i.e. settlement areas and roads were digitized from the maps which downloaded from www.banglarbhumi.gov.in website. The aforesaid model based susceptibility index map was intersected with the elements at risk data layers in Arc GIS 10.1 environment to get the different susceptibility values of settlement area and road length (Fig. 9.1). The susceptibility assessment was carried out on the whole Balason river basin as well as Mouza level. The prepared landslide susceptibility map using certainty factor model (Chap. 6) of the Balason river basin has been validated using ROC curve after Akgun et al. (2012), Bui et al. (2012) and Ozdemir and Altural (2013) and it is the graphical representation of the sensitivity and 1-specificity of landslide susceptibility models (Mohammady et al. 2012). The value of ROC curve called area under curve value (AUCV) which represents the predictive capability of the landslide susceptibility map. The ROC curve depicts 84.9% accuracy of the developed landslide susceptibility map of the Balson river basin (see Chap. 6, Sect. 6.4).

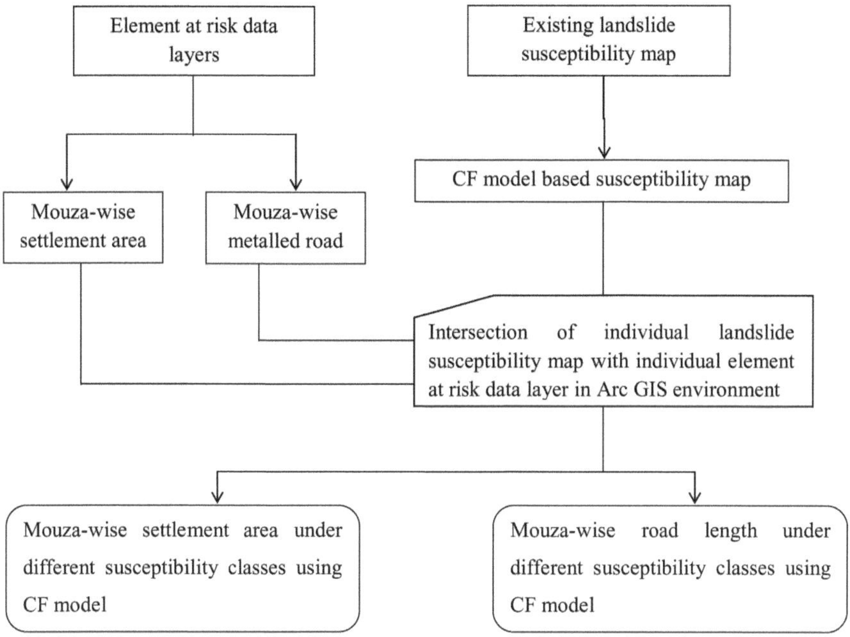

Fig. 9.1 Flow chart for susceptibility analysis and elements at risk assessment

9.3 Results and Discussion

9.3.1 Settlement Area Susceptibility

The total settlement area of the study area was 48,836,463.52 m^2 or nearly 48.84 km^2 out of which 31.50% of the area having very low susceptibility whereas 6.98% having low susceptibility, 16.78% area is under moderate susceptibility, 28.62% area is under high susceptibility and remaining. 16.13% area is under very high susceptibility zone based on CF Model (Fig. 9.2 and Table 9.1).

Dooteria Forest mouza was most vulnerable in terms of settlement area because 90.41% of settlement area fell under very high susceptibility zone using CF model but at the same time 0% area of that village was fall under very low and low susceptibility zone. Some other vulnerable villages which have more than 50% settlement area was under very high susceptibility zone were Pubong Tea Garden, Hill Cart Road, Rang Bul, Ghoom Pahar Forest, Jore Bunglow, Senchal Forest, Achhalal Hatta (Rong Bong Basty), Dhajea Khasmahal, Pussinbong Tea Garden and Chamu Tea Garden respectively. If we take the value of both very high and high landslide susceptibility zones the most vulnerable mouzas were Rang Bul (99.61%), Makai Bari Tea Garden (99.5%), Hill Cart Road (99.31%), Senchal Forest (99.26%), Pubong Tea Garden (98.96%), Manjua Forest (97.87%), Jore Bunglow (97.65%), Achhalal

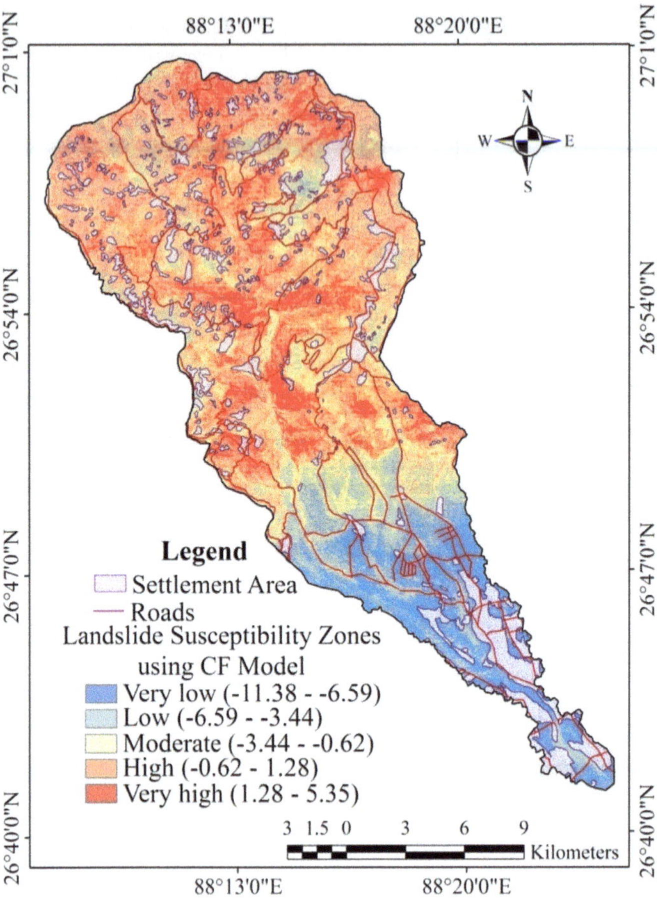

Fig. 9.2 Different susceptibility areas and elements at risk (settlement area and metalled road) using CF model

Table 9.1 Susceptibility analysis for settlement area using CF model of the whole Balason river basin

Susceptibility class	Absolute (m^2)	% of total settlement area
Very low susceptibility	15,381,588	31.50
Low susceptibility	3,407,664	6.98
Moderate susceptibility	8,193,236	16.78
High susceptibility	13,978,212	28.62
Very high susceptibility	7,875,764	16.13

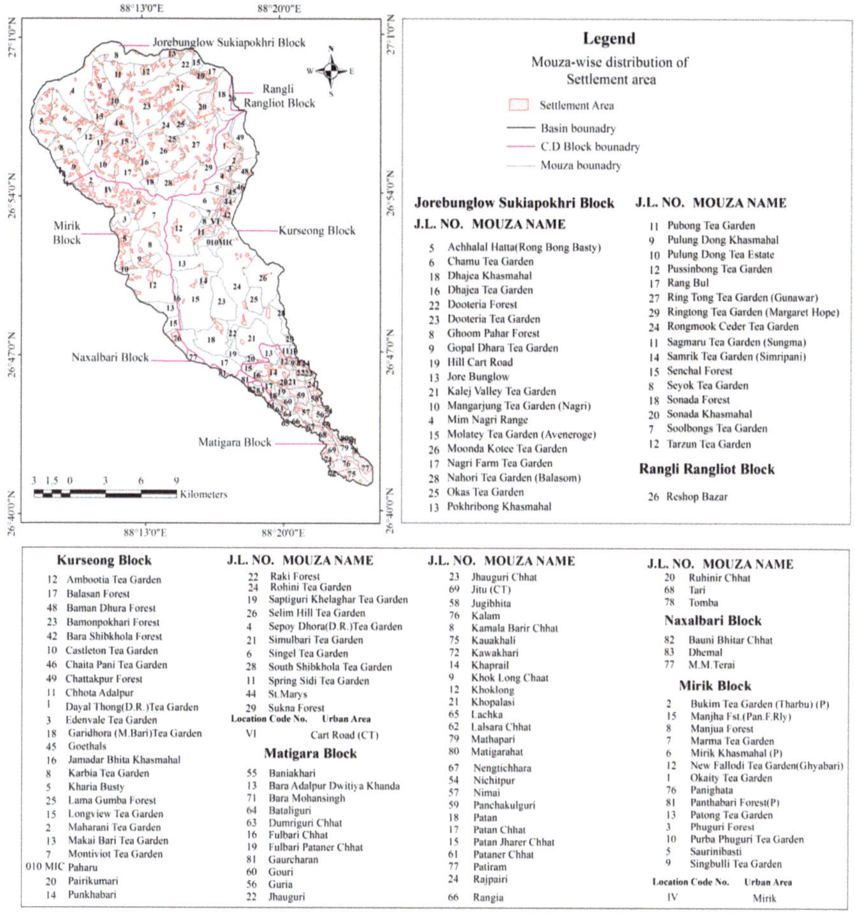

Fig. 9.3 Mouza wise distribution of settlement area within the river basin

Hatta (Rong Bong Basty) (97.63%), Dooteria Forest (97.62%), Chamu Tea Garden (96.87%), Dhajea Khasmahal (94.8%), Seyok Tea Garden (94.08%), Gopal Dhara Tea Garden (93.85%), Mirik Khasmahal (P) (92.87%), New Fallodi Tea Garden (Ghyabari) (92.72%) and Ringtong Tea Garden (Margaret Hope) (92.67%) respectively. The lowest vulnerable villages were Castleton Tea Garden, Chhota Adalpur, Nichitpur, Rangia and Bauni Bhitar Chhat because 100% of the settlement areas of the respective villages fall under the zone of very low susceptibility. Some other villages which have more than 75 percent area under very low susceptibility zones were Guria, Baniakhari, Kawakhari, Gaurcharan, Nengtichhara, Fulbari Chhat, Matigarahat, Tomba, Bara Mohansingh, Panchakulguri, Kamala Barir Chhat, Tari, Nimai, Khoklong, Jugibhita, Khok Long Chaat, Patiram, Fulbari Pataner Chhat, Patan Chhat, Kauakhali, Jhauguri, Mathapari, Rajpairi, Jhauguri Chhat, Khopalasi, Pataner Chhat, Patan and Bara Adalpur Dwitiya Khanda. The mouzas which have more than 70%

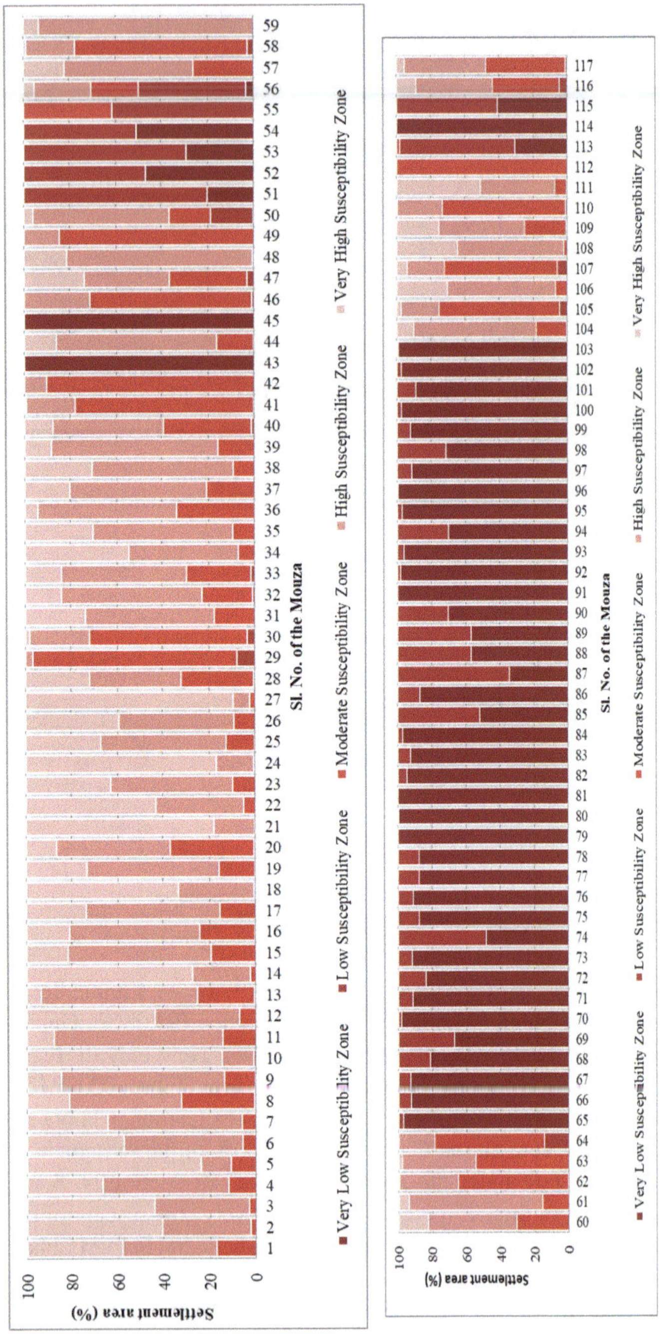

Fig. 9.4 Mouza wise settlement area falls under different susceptibility classes using FR model

area fall under moderate susceptibility were Manjha Fst. (Pan.F.Rly), Karbia Tea Garden, Rongmook Ceder Tea Garden, Punkhabari, Montiviot Tea Garden, Bara Shibkhola Forest, Purba Phuguri Tea Garden and Saurinibasti respectively (Figs. 9.3, 9.4 and Table 9.2).

9.3.2 Road Susceptibility

For the susceptibility assessment of metalled road of this basin CF model was incorporated and classified the area into five aforesaid susceptibility categories. The total road length was 308,892.5 m. CF model showed that about 30% of total roads length fall in between the very low to low susceptibility zones. About 17.75% roads were located on the very high susceptibility zone followed by high susceptibility zone (28.64%) and moderate susceptibility zone (22.86%) (see Fig. 9.2 and Table 9.3).

The most vulnerable villages among the entire study area in respect to metalled road was Rohini Tea Garden because 100% of total road was under very high susceptibility zone using CF model followed by Simulbari Tea Garden (91.19%), New Fallodi Tea Garden (Ghyabari) (77.49%), Ambootia Tea Garden (72.99%), Longview Tea Garden (70.8%), Ring Tong Tea Garden (Gunawar) (70.54%), Dooteria Tea Garden (69.63%), Bamonpokhari Forest (68.92%), Moonda Kotee Tea Garden (67.57%), Mirik (65.6%), Garidhora (M.Bari) Tea Garden (58.64%), Marma Tea Garden (57.25%) and Mim Nagri Range (50.22%) respectively. The lowest vulnerability in terms of roads was found in Rohini Tea Garden, Simulbari Tea Garden, New Fallodi Tea Garden (Ghyabari), Ambootia Tea Garden, Longview Tea Garden, Ring Tong Tea Garden (Gunawar), Dooteria Tea Garden, Bamonpokhari Forest and Moonda Kotee Tea Garden villages because entire road area under the very low susceptibility zones. More than 90% of the metalled road of Mirik, Garidhora (M.Bari) Tea Garden, Marma Tea Garden, Mim Nagri Range, Dhajea Khasmahal, Khaprail and Kalej Valley Tea Garden villages fall under the very low susceptibility class (Figs. 9.5, 9.6 and Table 9.4).

9.4 Conclusion

Landslide causative factors should be understood for reducing the number of casualties and damage to properties caused by landslides. The predictability of landslides susceptibility map depends on the scale, accuracy and completeness of landslide inventory map and various causative factors. In the present work, CF model based susceptibility maps was applied for the assessment of different susceptibility areas of settlement area and metalled road considering 15 landslide causative factors that is, elevation, aspect, slope, curvature, geology, soil, lineament density, distance to lineament, drainage density, distance to drainage, stream power index (SPI), topographic wetted index (TWI), rainfall (Ra), normalized differential vegetation index

Table 9.2 Mouza wise susceptibility analysis for settlement area using CF model

Sl. No.	JL No.	Name of the mouza	Mouza under the C.D block	Very low susceptibility area (%)	Low susceptibility area (%)	Moderate susceptibility area (%)	High susceptibility area (%)	Very high susceptibility area (%)	Settlement area (m^2)
1	4	Mim Nagri Range	Jorebunglow Sukiapokhri	0.00	0.00	17.38	41.04	41.58	166,160
2	5	Achhalal Hatta (Rong Bong Basty)	Jorebunglow Sukiapokhri	0.00	0.00	2.37	38.12	59.51	859,530.1
3	6	Chamu Tea Garden	Jorebunglow Sukiapokhri	0.00	0.00	3.14	40.82	56.05	753,659.5
4	7	Soolbongs Tea Garden	Jorebunglow Sukiapokhri	0.00	0.00	12.08	54.43	33.49	202,325.3
5	8	Ghoom Pahar Forest	Jorebunglow Sukiapokhri	0.00	0.00	10.98	12.96	76.06	88,088.32
6	8	Seyok Tea Garden	Jorebunglow Sukiapokhri	0.00	0.00	5.92	51.84	42.24	307,701.4
7	9	Gopal Dhara Tea Garden	Jorebunglow Sukiapokhri	0.00	0.00	6.15	58.33	35.52	620,011
8	10	Mangarjung Tea Garden (Nagri)	Jorebunglow Sukiapokhri	0.00	0.50	31.91	48.91	18.67	898,573.8
9	10	Pulung Dong Tea Estate	Jorebunglow Sukiapokhri	0.00	0.20	13.65	70.89	15.26	300,581.6
10	11	Pubong Tea Garden	Jorebunglow Sukiapokhri	0.00	0.00	1.04	13.88	85.08	447,047.3
11	11	Sagmaru Tea Garden (Sungma)	Jorebunglow Sukiapokhri	0.00	0.00	14.39	73.38	12.22	264,229.8

(continued)

Table 9.2 (continued)

Sl. No.	JL No.	Name of the mouza	Mouza under the C.D block	Very low susceptibility area (%)	Low susceptibility area (%)	Moderate susceptibility area (%)	High susceptibility area (%)	Very high susceptibility area (%)	Settlement area (m^2)
12	12	Pussinbong Tea Garden	Jorebunglow Sukiapokhri	0.00	0.00	7.38	36.30	56.32	772,419.4
13	12	Tarzun Tea Garden	Jorebunglow Sukiapokhri	0.00	0.72	24.68	68.41	6.18	261,135.3
14	13	Jore Bunglow	Jorebunglow Sukiapokhri	0.00	0.00	2.35	25.00	72.65	257,660.1
15	13	Pokhribong Khasmahal	Jorebunglow Sukiapokhri	0.00	0.06	19.33	62.14	18.47	440,365
16	14	Samrik Tea Garden (Simripani)	Jorebunglow Sukiapokhri	0.00	0.43	23.88	56.61	19.08	493,123.3
17	15	Molatey Tea Garden (Aveneroge)	Jorebunglow Sukiapokhri	0.00	0.00	15.61	57.86	26.53	523,646.3
18	15	Senchal Forest	Jorebunglow Sukiapokhri	0.00	0.00	0.74	32.42	66.84	111,818
19	16	Dhajea Tea Garden	Jorebunglow Sukiapokhri	0.00	0.27	15.74	57.33	26.65	451,068.2
20	17	Nagri Farm Tea Garden	Jorebunglow Sukiapokhri	0.00	0.70	35.92	49.96	13.43	647,492.9
21	17	Rang Bul	Jorebunglow Sukiapokhri	0.00	0.00	0.38	17.51	82.10	275,965
22	18	Dhajea Khasmahal	Jorebunglow Sukiapokhri	0.00	0.00	5.17	37.91	56.92	254,232.5

(continued)

Table 9.2 (continued)

Sl. No.	JL No.	Name of the mouza	Mouza under the C.D block	Very low susceptibility area (%)	Low susceptibility area (%)	Moderate susceptibility area (%)	High susceptibility area (%)	Very high susceptibility area (%)	Settlement area (m^2)
23	18	Sonada Forest	Jorebunglow Sukiapokhri	0.00	0.00	9.96	52.94	37.10	135,712.6
24	19	Hill Cart Road	Jorebunglow Sukiapokhri	0.00	0.00	0.69	16.41	82.90	123,200.3
25	20	Sonada Khasmahal	Jorebunglow Sukiapokhri	0.00	0.18	12.48	54.26	33.08	2,293,492
26	21	Kalej Valley Tea Garden	Jorebunglow Sukiapokhri	0.00	0.00	9.45	49.94	40.60	859,543.5
27	22	Dooteria Forest	Jorebunglow Sukiapokhri	0.00	0.00	2.38	7.21	90.41	35,680.49
28	23	Dooteria Tea Garden	Jorebunglow Sukiapokhri	0.00	0.71	31.25	39.82	28.22	1,406,970
29	24	Rongmook Ceder Tea Garden (CT)	Jorebunglow Sukiapokhri	0.00	8.10	88.76	3.14	0.00	351,389.8
30	25	Okas Tea Garden	Jorebunglow Sukiapokhri	0.00	3.45	68.48	26.47	1.60	438,865.9
31	26	Moonda Kotee Tea Garden	Jorebunglow Sukiapokhri	0.00	0.00	17.82	55.83	26.34	927,060.3
32	27	Ring Tong Tea Garden (Gunawar)	Jorebunglow Sukiapokhri	0.00	0.91	22.05	60.88	16.16	662,797.5
33	28	Nahori Tea Garden (Balasom)	Jorebunglow Sukiapokhri	0.00	1.58	27.92	54.61	15.89	365,427.6

(continued)

Table 9.2 (continued)

Sl. No.	JL No.	Name of the mouza	Mouza under the C.D block	Very low susceptibility area (%)	Low susceptibility area (%)	Moderate susceptibility area (%)	High susceptibility area (%)	Very high susceptibility area (%)	Settlement area (m^2)
34	29	Ringtong Tea Garden (Margaret Hope)	Jorebunglow Sukiapokhri	0.00	0.00	7.33	47.58	45.09	279,451.1
35	1	Dayal Thong (D.R.) Tea Garden	Kurseong	0.00	0.00	9.69	60.39	29.91	599,701
36	2	Maharani Tea Garden	Kurseong	0.00	0.00	33.80	60.45	5.75	376,207.9
37	3	Edenvale Tea Garden	Kurseong	0.00	0.00	20.73	59.59	19.68	203,328.7
38	4	Sepoy Dhora (D.R.) Tea Garden	Kurseong	0.00	0.00	9.36	61.27	29.36	331,943.1
39	5	Kharia Busty	Kurseong	0.00	0.00	15.99	72.05	11.97	244,917.8
40	6	Singel Tea Garden	Kurseong	0.00	1.25	37.80	48.52	12.43	338,490.3
41	7	Montiviot Tea Garden	Kurseong	0.00	0.25	77.58	21.39	0.78	248,953.7
42	8	Karbia Tea Garden	Kurseong	0.00	0.11	90.12	9.78	0.00	33,694.44
43	10	Castleton Tea Garden	Kurseong	100.00	0.00	0.00	0.00	0.00	36.80781
44	10	Paharu	Kurseong	0.00	0.20	16.00	69.62	14.18	218,813.1
45	11	Chhota Adalpur	Kurseong	100.00	0.00	0.00	0.00	0.00	19,290.38

(continued)

Table 9.2 (continued)

Sl. No.	JL No.	Name of the mouza	Mouza under the C.D block	Very low susceptibility area (%)	Low susceptibility area (%)	Moderate susceptibility area (%)	High susceptibility area (%)	Very high susceptibility area (%)	Settlement area (m^2)
46	11	Spring Side Tea Garden	Kurseong	0.00	1.17	69.99	28.65	0.19	478,027.4
47	12	Ambootia Tea Garden	Kurseong	0.00	2.96	33.39	37.26	26.38	403,992.3
48	13	Makai Bari Tea Garden	Kurseong	0.00	0.00	0.50	80.57	18.93	19,715.87
49	14	Punkhabari	Kurseong	0.00	0.00	84.24	15.76	0.00	5609.832
50	15	Longview Tea Garden	Kurseong	0.26	18.45	17.67	59.79	3.83	263,851.4
51	17	Balasan Forest	Kurseong	19.99	80.01	0.00	0.00	0.00	1643.251
52	18	Garidhora (M.Bari) Tea Garden	Kurseong	46.91	53.09	0.00	0.00	0.00	364,506.2
53	20	Pairikumari	Kurseong	29.01	70.99	0.00	0.00	0.00	15,695.25
54	21	Simulbari Tea Garden	Kurseong	51.08	48.92	0.00	0.00	0.00	156,060.3
55	23	Bamonpokhari Forest	Kurseong	0.00	61.49	38.51	0.00	0.00	1044.401
56	24	Rohini Tea Garden	Kurseong	3.29	46.53	20.75	24.87	4.56	813,801.9
57	26	Selim Hill Tea Garden	Kurseong	0.00	0.00	25.98	56.02	18.01	77,241.37
58	42	Bara Shibkhola Forest	Kurseong	0.00	2.91	74.61	21.27	1.20	371,497.2

(continued)

Table 9.2 (continued)

Sl. No.	JL No.	Name of the mouza	Mouza under the C.D block	Very low susceptibility area (%)	Low susceptibility area (%)	Moderate susceptibility area (%)	High susceptibility area (%)	Very high susceptibility area (%)	Settlement area (m^2)
59	44	St. Marys	Kurseong	0.00	0.00	0.00	93.33	6.67	4082.219
60	45	Goethals	Kurseong	0.00	0.00	30.80	52.02	17.18	3740.112
61	46	Chaita Pani Tea Garden	Kurseong	0.00	0.00	15.61	78.47	5.92	27,062.19
62	48	Baman Dhura Forest	Kurseong	0.00	0.30	64.49	34.85	0.36	158,531.7
63	49	Chattakpur Forest	Kurseong	0.00	0.00	54.84	43.03	2.13	26,607.52
64	65	Lama Gumba Forest	Kurseong	0.00	14.38	64.34	21.28	0.00	40,831.59
65	8	Kamala Barir Chhat	Matigara	97.02	2.98	0.00	0.00	0.00	327,565.7
66	9	Khok Long Chaat	Matigara	92.42	7.58	0.00	0.00	0.00	128,706.7
67	12	Khoklong	Matigara	93.02	6.98	0.00	0.00	0.00	538,384.6
68	13	Bara Adalpur Dwitiya Khanda	Matigara	81.06	18.94	0.00	0.00	0.00	140,402.1
69	14	Khaprail	Matigara	67.21	32.78	0.02	0.00	0.00	100,497.2
70	16	Fulbari Chhat	Matigara	98.09	1.91	0.00	0.00	0.00	133,566.3
71	17	Patan Chhat	Matigara	91.66	8.34	0.00	0.00	0.00	16,441.42
72	18	Patan	Matigara	83.79	16.21	0.00	0.00	0.00	90,801.11
73	19	Fulbari Pataner Chhat	Matigara	91.76	8.24	0.00	0.00	0.00	77,999.2
74	20	Ruhinir Chhat	Matigara	48.24	51.76	0.00	0.00	0.00	1,271,384
75	21	Khopalasi	Matigara	87.45	12.55	0.00	0.00	0.00	496,119.2

(continued)

Table 9.2 (continued)

Sl. No.	JL No.	Name of the mouza	Mouza under the C.D block	Very low susceptibility area (%)	Low susceptibility area (%)	Moderate susceptibility area (%)	High susceptibility area (%)	Very high susceptibility area (%)	Settlement area (m^2)
76	22	Jhauguri	Matigara	91.21	8.79	0.00	0.00	0.00	129,834.7
77	23	Jhauguri Chhat	Matigara	87.47	12.53	0.00	0.00	0.00	587,316.8
78	24	Rajpairi	Matigara	87.70	12.30	0.00	0.00	0.00	617,540.5
79	54	Nichitpur	Matigara	100.00	0.00	0.00	0.00	0.00	1,267,571
80	55	Baniakhari	Matigara	99.74	0.26	0.00	0.00	0.00	163,955.5
81	56	Guria	Matigara	99.95	0.05	0.00	0.00	0.00	415,445
82	57	Nimai	Matigara	94.80	5.20	0.00	0.00	0.00	42,080.21
83	58	Jugibhita	Matigara	92.70	7.30	0.00	0.00	0.00	877,022.4
84	59	Panchakulguri	Matigara	97.05	2.95	0.00	0.00	0.00	79,257.55
85	60	Gouri	Matigara	52.02	47.98	0.00	0.00	0.00	739,534.3
86	61	Pataner Chhat	Matigara	87.01	12.99	0.00	0.00	0.00	1,494,400
87	62	Lalsara Chhat	Matigara	34.51	65.49	0.00	0.00	0.00	15,725.8
88	63	Dumriguri Chhat	Matigara	56.92	43.08	0.00	0.00	0.00	38,877.54
89	64	Bataliguri	Matigara	56.97	43.03	0.00	0.00	0.00	513,290.8
90	65	Lachka	Matigara	70.16	29.84	0.00	0.00	0.00	76,108.15
91	66	Rangia	Matigara	100.00	0.00	0.00	0.00	0.00	260,004.9
92	67	Nengtichhara	Matigara	98.25	1.75	0.00	0.00	0.00	678.8088
93	68	Tari (CT)	Matigara	96.25	3.75	0.00	0.00	0.00	1,045,392
94	69	Jitu	Matigara	69.86	30.14	0.00	0.00	0.00	1,524,853

(continued)

Table 9.2 (continued)

Sl. No.	JL No.	Name of the mouza	Mouza under the C.D block	Very low susceptibility area (%)	Low susceptibility area (%)	Moderate susceptibility area (%)	High susceptibility area (%)	Very high susceptibility area (%)	Settlement area (m^2)
95	71	Bara Mohansingh	Matigara	97.18	2.82	0.00	0.00	0.00	148,871.3
96	72	Kawakhari	Matigara	99.70	0.30	0.00	0.00	0.00	360,667.4
97	75	Kauakhali	Matigara	91.49	8.51	0.00	0.00	0.00	294,517
98	76	Kalam	Matigara	71.32	28.68	0.00	0.00	0.00	1,265,543
99	77	Patiram	Matigara	92.19	7.81	0.00	0.00	0.00	702,726.9
100	78	Tomba	Matigara	97.41	2.59	0.00	0.00	0.00	163,834.2
101	79	Mathapari	Matigara	89.00	11.00	0.00	0.00	0.00	147,375.4
102	80	Matigarahat	Matigara	97.53	2.47	0.00	0.00	0.00	938,457.9
103	81	Gaurcharan	Matigara	99.12	0.88	0.00	0.00	0.00	32,290.67
104	2	Bukim Tea Garden (Tharbu) (P)	Mirik	0.00	0.23	18.05	71.74	9.98	268,663.6
105	5	Saurinibasti	Mirik	0.00	4.41	70.71	22.25	2.63	1,171,687
106	6	Mirik Khasmahal (P)	Mirik	0.00	0.00	7.13	63.19	29.68	611,180.3
107	7	Marma Tea Garden	Mirik	0.00	5.98	66.15	21.87	6.00	1,044,318
108	8	Manjua Forest	Mirik	0.00	0.00	2.13	62.09	35.78	115,860.5
109	9	Singbulli Tea Garden	Mirik	0.00	0.57	24.37	50.42	24.63	638,422.6
110	10	Purba Phuguri Tea Garden	Mirik	0.00	1.02	71.94	26.54	0.49	179,249.8

(continued)

Table 9.2 (continued)

Sl. No.	JL No.	Name of the mouza	Mouza under the C.D block	Very low susceptibility area (%)	Low susceptibility area (%)	Moderate susceptibility area (%)	High susceptibility area (%)	Very high susceptibility area (%)	Settlement area (m^2)
111	12	New Fallodi Tea Garden (Gayabari)	Mirik	0.00	0.08	7.21	43.49	49.23	808,155.7
112	15	Manjha Fst. (Pan.F.Rly)	Mirik	0.00	0.00	99.88	0.12	0.00	9706.895
113	76	Panighata	Mirik	30.60	67.73	1.66	0.00	0.00	375,201.1
114	82	Bauni Bhitar Chhat	Naxalbari	100.00	0.00	0.00	0.00	0.00	1476.317
115	83	Dhemal	Naxalbari	40.80	59.20	0.00	0.00	0.00	57,315.13
	Location code	Urban area	Mouza under the C.D block	Very low susceptibility area (%)	Low susceptibility area (%)	Moderate susceptibility area (%)	High susceptibility area (%)	Very high susceptibility area (%)	Settlement area (m^2)
116	IV	Mirik (CT)	Mirik	0.00	4.57	39.33	44.85	11.25	2,015,834
117	VI	Cart Road (CT)	Kurseong	0.00	0.87	46.90	47.67	4.55	1,149,008

Table 9.3 Susceptibility analysis for metalled road using CF model

Susceptibility class	Absolute (m)	% of total road
Very low susceptibility	55,918.98	18.10
Low susceptibility	39,672.15	12.84
Moderate susceptibility	70,001.01	22.66
High susceptibility	88,460.68	28.64
Very high susceptibility	54,839.72	17.75

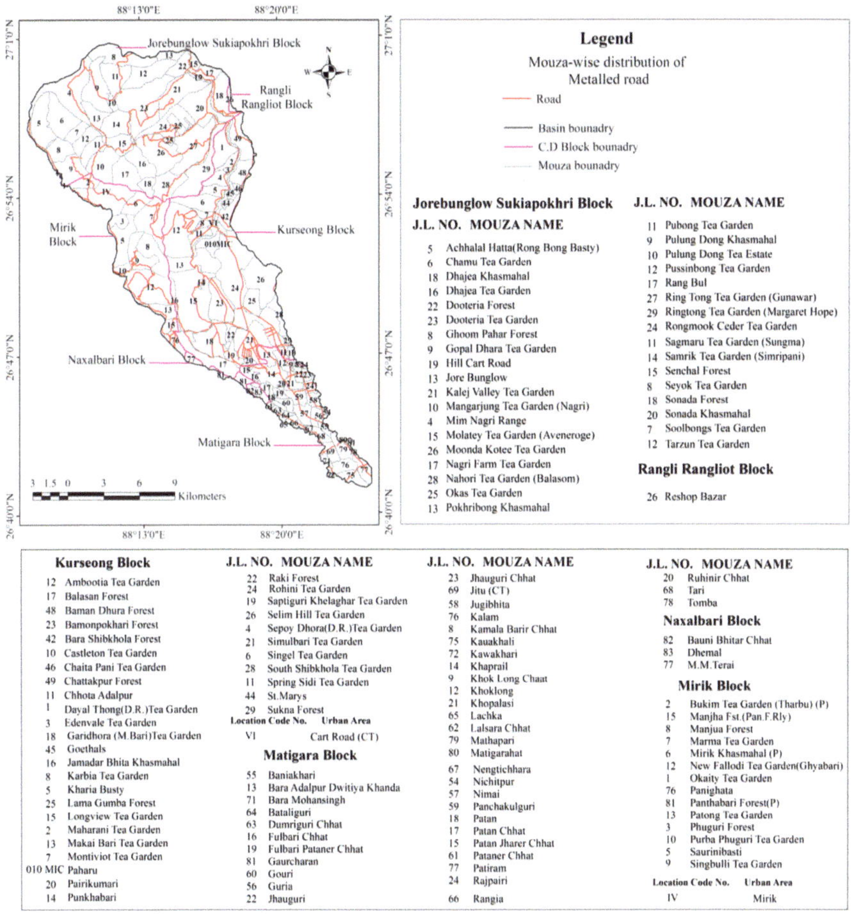

Fig. 9.5 Mouza wise metalled road within the river basin

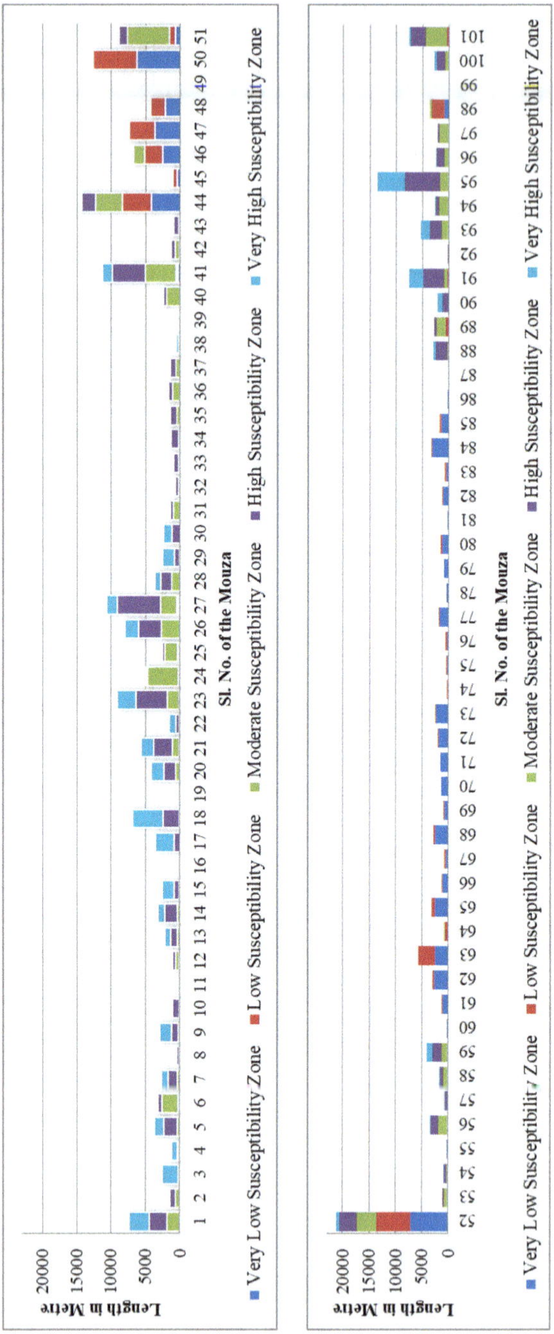

Fig. 9.6 Mouza wise road length falls under different susceptibility classes using LR model

Table 9.4 Mouza wise susceptibility analysis for metalled road using CF model

Sl. No.	Jl. No.	Name of the mouza	Mouza under the C.D block	Very low susceptibility (m)	Low susceptibility (m)	Moderate susceptibility (m)	High susceptibility (m)	Very high susceptibility (m)	Total road length (m)
1	4	Mim Nagri Range	Jorebunglow Sukiapokhri	0.00	0.00	1776.404118	2608.047541	2894.34829	7278.799949
2	5	Achhalal Hatta (Rong Bong Basty)	Jorebunglow Sukiapokhri	0.00	0.00	627.876921	767.4166	133.423318	1528.716839
3	8	Ghoom Pahar Forest	Jorebunglow Sukiapokhri	0.00	0.00	0.00	229.877831	2378.205532	2608.083363
4	8	Seyok Tea Garden	Jorebunglow Sukiapokhri	0.00	0.00	68.105095	305.196788	893.765016	1267.066899
5	9	Gopal Dhara Tea Garden	Jorebunglow Sukiapokhri	0.00	0.00	308.737786	2038.981652	1301.121918	3648.841356
6	9	Pulung Dong Khasmahal	Jorebunglow Sukiapokhri	0.00	136.382371	2320.970393	589.980786	243.177229	3290.510779
7	10	Mangarjung Tea Garden (Nagri)	Jorebunglow Sukiapokhri	0.00	0.00	371.142627	1367.283892	959.072519	2697.499038
8	10	Pulung Dong Tea Estate	Jorebunglow Sukiapokhri	0.00	0.00	84.656265	365.663703	177.707407	628.027375
9	11	Pubong Tea Garden	Jorebunglow Sukiapokhri	0.00	0.00	197.485314	1010.110123	1711.962042	2919.557479
10	11	Sagmaru Tea Garden (Sungma)	Jorebunglow Sukiapokhri	0.00	0.00	46.645537	999.941703	277.721746	1324.308986
11	12	Tarzun Tea Garden	Jorebunglow Sukiapokhri	0.00	0.00	0.00	5.2171	0.00	5.2171

(continued)

Table 9.4 (continued)

Sl. No.	Jl. No.	Name of the mouza	Mouza under the C.D block	Very low susceptibility (m)	Low susceptibility (m)	Moderate susceptibility (m)	High susceptibility (m)	Very high susceptibility (m)	Total road length (m)
12	13	Pokhribong Khasmahal	Jorebunglow Sukiapokhri	0.00	34.672983	542.372638	498.341819	0.00	1075.38744
13	14	Samrik Tea Garden (Simripani)	Jorebunglow Sukiapokhri	0.00	0.00	377.678396	1009.115627	835.9512	2222.745223
14	15	Molatey Tea Garden (Avenerog)	Jorebunglow Sukiapokhri	0.00	0.00	331.291102	1795.909818	1004.999237	3132.200157
15	15	Senchal Forest	Jorebunglow Sukiapokhri	0.00	0.00	49.627996	761.074658	1797.331122	2608.033776
16	16	Dhajea Tea Garden	Jorebunglow Sukiapokhri	0.00	0.00	0.00	105.485434	219.817931	325.303365
17	17	Rang Bul	Jorebunglow Sukiapokhri	0.00	0.00	25.66015	774.053387	2753.083644	3552.797181
18	18	Dhajea Khasmahal	Jorebunglow Sukiapokhri	0.00	0.00	159.279892	2212.668711	4522.495708	6894.444311
19	19	Hill Cart Road	Jorebunglow Sukiapokhri	0.00	0.00	0.00	96.995036	262.110173	359.105209
20	20	Sonada Khasmahal	Jorebunglow Sukiapokhri	0.00	0.00	616.651246	1728.696929	1768.459695	4113.80787
21	21	Kalej Valley Tea Garden	Jorebunglow Sukiapokhri	0.00	27.320953	1058.44311	2680.336419	1800.365592	5566.466074
22	22	Dooteria Forest	Jorebunglow Sukiapokhri	0.00	0.00	74.005929	581.681362	877.973733	1533.661024

(continued)

Table 9.4 (continued)

Sl. No.	Jl. No.	Name of the mouza	Mouza under the C.D block	Very low susceptibility (m)	Low susceptibility (m)	Moderate susceptibility (m)	High susceptibility (m)	Very high susceptibility (m)	Total road length (m)
23	23	Dooteria Tea Garden	Jorebunglow Sukiapokhri	0.00	44.846578	1673.227995	4545.664304	2699.209258	8962.948135
24	24	Rongmook Ceder Tea Garden	Jorebunglow Sukiapokhri	0.00	120.162571	4395.453694	156.166954	81.294433	4753.077652
25	25	Okas Tea Garden	Jorebunglow Sukiapokhri	0.00	158.543208	1888.424238	327.559165		2374.526611
26	26	Moonda Kotee Tea Garden	Jorebunglow Sukiapokhri	0.00	22.135839	2593.396356	3396.56757	1879.098557	7891.198322
27	27	Ring Tong Tea Garden (Gunawar)	Jorebunglow Sukiapokhri	0.00	221.36663	2385.864284	6168.194135	1629.178586	10,404.60364
28	28	Nahori Tea Garden (Balasom)	Jorebunglow Sukiapokhri	0.00	7.793841	1146.96792	1630.584964	865.167077	3650.513802
29	29	Ringtong Tea Garden (Margaret Hope)	Jorebunglow Sukiapokhri	0.00	0.00	49.693422	733.010647	1794.745739	2577.449808
30	1	Dayal Thong (D.R.) Tea Garden	Kurseong	0.00	0.00	11.40188	1186.628769	1208.708815	2406.739464
31	2	Maharani Tea Garden	Kurseong	0.00	0.00	907.464649	593.873848	111.193707	1612.532204
32	3	Edenvale Tea Garden	Kurseong	0.00	0.00	241.89503	431.698836	9.472583	683.066449

(continued)

Table 9.4 (continued)

Sl. No.	Jl. No.	Name of the mouza	Mouza under the C.D block	Very low susceptibility (m)	Low susceptibility (m)	Moderate susceptibility (m)	High susceptibility (m)	Very high susceptibility (m)	Total road length (m)
33	4	Sepoy Dhora (D.R.) Tea Garden	Kurseong	0.00	0.00	172.128338	768.977187	299.950317	1241.055842
34	5	Kharia Busty	Kurseong	0.00	0.00	190.462635	1113.211757	238.008512	1541.682904
35	6	Singel Tea Garden	Kurseong	0.00	0.00	527.43312	898.508259	180.234918	1606.176297
36	7	Montiviot Tea Garden	Kurseong	0.00	0.00	1142.96692	598.776738	70.450629	1812.194287
37	8	Karbia Tea Garden	Kurseong	0.00	0.00	557.948819	930.745196	9.809899	1498.503914
38	010 MIC	Paharu	Kurseong	0.00	0.00	0.00	175.331117	425.135222	600.466339
39	11	Chhota Adalpur	Kurseong	1972.263089	0.00	0.00	0.00	0.00	1972.263089
40	11	Spring Sidi Tea Garden	Kurseong	0.00	22.763794	1866.610769	574.259767	20.882431	2484.516761
41	12	Ambootia Tea Garden	Kurseong	0.00	293.074274	4468.769293	4699.217986	1427.537452	10,888.59901
42	13	Makai Bari Tea Garden	Kurseong	0.00	33.304526	622.844508	692.464416	253.19585	1601.8093
43	14	Punkhabari	Kurseong	0.00	0.00	242.979178	748.348518	32.117283	1023.444979
44	15	Longview Tea Garden	Kurseong	29.506106	4181.441001	3874.789619	2096.028032	268.047761	10,449.81252
45	16	Jamadar Bhita Khasmahal	Kurseong	0.00	534.014251	267.237669	0.00	0.00	801.25192

(continued)

Table 9.4 (continued)

Sl. No.	Jl. No.	Name of the mouza	Mouza under the C.D block	Very low susceptibility (m)	Low susceptibility (m)	Moderate susceptibility (m)	High susceptibility (m)	Very high susceptibility (m)	Total road length (m)
46	17	Balasan Forest	Kurseong	135.733494	2575.743374	1550.439727	43.374942	0.00	4305.291537
47	18	Garidhora (M.Bari) Tea Garden	Kurseong	3751.246562	3683.14762	156.125308	0.00	0.00	7590.51949
48	19	Saptiguri Khelaghar Tea Garden	Kurseong	227.761505	2162.682975	47.563824	0.00	0.00	2438.008304
49	20	Pairikumari	Kurseong	151.803241	153.617641	0.00	0.00	0.00	305.420882
50	21	Simulbari Tea Garden	Kurseong	8771.672892	6284.345572	0.00	0.00	0.00	15,056.01846
51	23	Bamonpokhari Forest	Kurseong	0.00	774.821829	6025.467685	1251.861455	25.037337	8077.188306
52	24	Rohini Tea Garden	Kurseong	7105.71194	6477.487343	3758.268604	3304.346925	668.827634	21,314.64245
53	25	Lama Gumba Forest	Kurseong	0.00	129.309668	565.172142	333.554737	0.00	1028.036547
54	26	Selim Hill Tea Garden	Kurseong	0.00	0.00	261.522096	480.999904	81.690104	824.212104
55	29	Sukna Forest	Kurseong	209.673526	0.00	0.00	0.00	0.00	209.673526
56	42	Bara Shibkhola Forest	Kurseong	0.00	95.351165	1699.572179	1547.110065	126.984762	3469.018171
57	46	Chaita Pani Tea Garden	Kurseong	0.00	0.00	152.93243	395.298875	28.56523	576.796535
58	48	Baman Dhura Forest	Kurseong	0.00	0.00	889.227398	709.254534	91.35545	1689.837382
59	49	Chattakpur Forest	Kurseong	0.00	13.060086	1164.79086	1885.185646	1053.308991	4116.345583

(continued)

Table 9.4 (continued)

Sl. No.	Jl. No.	Name of the mouza	Mouza under the C.D block	Very low susceptibility (m)	Low susceptibility (m)	Moderate susceptibility (m)	High susceptibility (m)	Very high susceptibility (m)	Total road length (m)
60	9	Khok Long Chaat	Matigara	288.065461	0.00	0.00	0.00	0.00	288.065461
61	12	Khoklong	Matigara	1025.782805	30.731397	0.00	0.00	0.00	1056.514202
62	13	Bara Adalpur Dwitiya Khanda	Matigara	2619.213611	438.403268	0.00	0.00	0.00	3057.616879
63	14	Khaprail	Matigara	2451.434562	3187.904013	0.00	0.00	0.00	5639.338575
64	15	Patan Jharer Chhat	Matigara	29.866969	599.908412	39.183401	0.00	0.00	668.958782
65	21	Khopalasi	Matigara	2452.2563	652.58175	0.00	0.00	0.00	3104.83805
66	22	Jhauguri	Matigara	1034.816182	153.498332	0.00	0.00	0.00	1188.314514
67	23	Jhauguri Chhat	Matigara	508.274208	290.156154	0.00	0.00	0.00	798.430362
68	24	Rajpairi	Matigara	2528.582463	260.538953	0.00	0.00	0.00	2789.121416
69	55	Baniakhar	Matigara	784.885068	29.670668	0.00	0.00	0.00	814.555736
70	56	Guria	Matigara	1455.411935	0.00	0.00	0.00	0.00	1455.411935
71	57	Nimai	Matigara	1590.024096	0.00	0.00	0.00	0.00	1590.024096
72	58	Jugibhita	Matigara	1835.995084	17.073328	0.00	0.00	0.00	1853.068412
73	59	Panchakulguri	Matigara	2301.64049	130.318323	0.00	0.00	0.00	2431.958813
74	62	Lalsara Chhat	Matigara	71.297849	129.367171	0.00	0.00	0.00	200.66502
75	63	Dumriguri Chhat	Matigara	321.948598	191.243211	0.00	0.00	0.00	513.191809
76	64	Bataliguri	Matigara	364.359605	264.669501	0.00	0.00	0.00	629.029106
77	66	Rangia	Matigara	1783.690656	32.263428	0.00	0.00	0.00	1815.954084

(continued)

Table 9.4 (continued)

Sl. No.	Jl. No.	Name of the mouza	Mouza under the C.D block	Very low susceptibility (m)	Low susceptibility (m)	Moderate susceptibility (m)	High susceptibility (m)	Very high susceptibility (m)	Total road length (m)
78	67	Nengtichhara	Matigara	423.996334	0.00	0.00	0.00	0.00	423.996334
79	68	Tari	Matigara	911.080303	0.00	0.00	0.00	0.00	911.080303
80	69	Jitu	Matigara	1211.983287	381.846502	0.00	0.00	0.00	1593.829789
81	71	Bara Mohansingh	Matigara	261.243083	0.00	0.00	0.00	0.00	261.243083
82	75	Kauakhali	Matigara	1071.960782	122.996731	0.00	0.00	0.00	1194.957513
83	76	Kalam	Matigara	489.466572	263.777066	0.00	0.00	0.00	753.243638
84	77	Patiram	Matigara	3147.072307	78.166223	0.00	0.00	0.00	3225.23853
85	79	Mathapari	Matigara	1418.202649	236.395867	0.00	0.00	0.00	1654.598516
86	80	Matigarahat	Matigara	231.354448	0.00	0.00	0.00	0.00	231.354448
87	1	Okaity Tea Garden	Mirik	0.00	0.00	3.51416	0.00	0.00	3.51416
88	2	Bukim Tea Garden (Tharbu) (P)	Mirik	0.00	0.00	248.409146	2220.47777	526.237683	2995.124599
89	5	Saurinibasti	Mirik	0.00	556.819094	1756.700144	510.314231	0.00	2823.833469
90	6	Mirik Khasmahal (P)	Mirik	0.00	0.00	126.196566	1075.686917	960.578186	2162.461669
91	7	Marma Tea Garden	Mirik	0.00	216.829196	776.721923	3874.899374	2719.496341	7587.946834
92	8	Manjua Forest	Mirik	0.00	0.00	75.856939	111.10563	0.00	186.962569
93	9	Singbulli Tea Garden	Mirik	0.00	0.00	1359.392043	2191.585663	1877.210671	5428.188377
94	10	Purba Phuguri Tea Garden	Mirik	0.00	138.443069	1699.005958	728.565077	1.917623	2567.931727

(continued)

Table 9.4 (continued)

Sl. No.	Jl. No.	Name of the mouza	Mouza under the C.D block	Very low susceptibility (m)	Low susceptibility (m)	Moderate susceptibility (m)	High susceptibility (m)	Very high susceptibility (m)	Total road length (m)
95	12	New Fallodi Tea Garden (Chyabari)	Mirik	0.00	77.637791	1674.117106	6643.412002	5237.639076	13,632.80598
96	13	Patong Tea Garden	Mirik	0.00	0.00	986.214458	1288.813968	238.289162	2513.317588
97	15	Manjha Fst. (Pan.F.Rly.)	Mirik	0.00	189.224948	1692.08056	344.933654	0.00	2226.239162
98	76	Panighata	Mirik	947.649419	2390.255799	338.798427	0.00	0.00	3676.703645
99	26	Reshop Bazar	Rangli Rangliot	0.00	0.00	0.00	0.00	84.656274	84.656274
	Location code	Name of urban units	Urban units under block	Very low susceptibility (m)	Low susceptibility (m)	Moderate susceptibility (m)	High susceptibility (m)	Very high susceptibility (m)	Total road length (m)
100	VI	Cart Road (CT)	Kurseong	0.00	18.989519	665.794743	1613.062865	539.614317	2837.461444
101	IV	Mirik	Mirik	0.00	403.090252	4019.447334	2919.152298	259.077126	7600.76701

(NDVI) and land use and land cover (LULC). Very high susceptibility zones for settlement area cover 16.13% to the total settlement area of the Balason river basin. Similarly, 17.75% metalled road to the total metalled road fall in the very high susceptibility categories. The most vulnerable mouza in respect of settlement area and metalled road was Dooteria Forest mouza and Dhajea Tea Garden and Rohini Tea Garden mouza based on the CF model. The findings of the research can be utilized to prioritize risk management so that the losses and damages of properties and lives can be minimized due to future landslides event. In addition, the results can help the planners, developers and engineers for reducing the hazardous impact of landslides on road network and human settlement area.

References

Akgun A, Sezer EA, Nefesliogl HA, Gokceoglu C, Pradhan B (2012) An easy-to-use MATLAB program (MamLand) for the assessment of landslide susceptibility using a Mamdani fuzzy algorithm. Comput Geosci 38:23–34

Bhandari RK (2007) The Indian landslide scenario: strategic issues and action points. Disaster Dev 1:155–168

Bhattacharya SK (1998) The assessment of forest depletion and its conservation measures of the Rakti basin in Darjeeling. In: De AK et al (eds) Sustainable development and environment. Cosmo Publications, New Delhi, pp 257–273

Bhattacharya SK (1999) A constructive approach to landslides through susceptibility zoning and case study in the Rakti basin of Eastern Himalaya. Trans Jpn Geomorphol Union 20(3):317–333

Bhattacharya SK (2002) Tindharia Cricket Colony landslide in Darjeeling Hills—a comparative account, 1996 and 2000. In: Basu S (ed) Changing environmental scenario of the Indian subcontinent. ACB Publication, Calcutta, pp 400–405

Bhattacharya SK (2010) The study of some Aila induced landslides in the Darjeeling Hills, West Bengal, India. In: Sarkar S (ed) Geo-hazards in sub-Himalayan North Bengal, SAP-DRS I (monograph 1). Registrar, University of North Bengal, Rajarammohanpur, pp 70–75

Bui DT, Pradhan B, Lofman O, Revhaug I, Dick OB (2012) Spatial prediction of landslide hazards in Hoa Binh province (Vietnam): a comparative assessment of the efficacy of evidential belief functions and fuzzy logic models. Catena 96:28–40

Carrara A, Cardinali M, Guzzetti F, Reichenbach P (1995) GIS technology in mapping landslide hazard. In: Geographical information systems in assessing natural hazards. Kluwer Academic Publishers, Dordrecht, pp 135–175

Ercanoglu M (2005) Landslide susceptibility assessment of SE Bartin (West Black Sea region, Turkey) by artificial neural networks. Nat Hazards Earth Syst Sci 5:979–992

Ermini L, Catani F, Casagli N (2005) Artificial neural networks applied to landslide susceptibility assessment. Geomorphology 66:327–343

Fell R, Corominas J, Bonnard C, Cascini L, Leroi E, Savage WZ (2008) Guidelines for landslide susceptibility, hazard and risk zoning for land use planning. Eng Geol 102:85–98

Hays WW (1981) Facing geologic and hydrologic hazards: earth-science considerations. Professional paper, 1240-B. U.S. Geological Survey, Reston, Virginia

Mandal S, Maiti R (2014) Semi-quantitative approaches for landslide assessment and prediction. Springer

Mandal B, Mandal S (2017) Landslide susceptibility mapping using modified information value model in the Lish river basin of Darjiling Himalaya. Spat Inf Res 25(2):205–218. https://doi.org/10.1007/s41324-017-0096-4

Mohammady M, Pourghasemi HR, Pradhan B (2012) Landslide susceptibility mapping at Golestan Province, Iran: a comparison between frequency ratio, Dempster–Shafer, and weights-of-evidence models. J Asian Earth Sci 61:221–236. https://doi.org/10.1016/j.jseaes.2012.10.005

Nautiyal SP (1951) A geological report on the hill slope stability in and around Darjeeling, W.B. Unpublished report of the geological survey of India

Ozdemir A, Altural T (2013) A comparative study of frequency ratio, weights of evidence and logistic regression methods for landslide susceptibility mapping: Sultan Mountains, SW Turkey. J Asian Earth Sci 64:180–197

Sarkar S, Kanungo DP (2010) Landslides in relation to terrain parameters—a remote sensing and GIS approach. GIS development.net. Nat Hazard Manag 1–6

Sengupta CK (1995) Detailed study of geofactors in selected hazard prone stretches along the surface communication routes in parts of Darjeeling and Sikkim Himalaya, Phase-I, Part-I (Rongtong-Kurseong road section). Unpublished GSI report, FS 1993–94 & 1994–95

Sinha BN (1974) Inspection of the alternative road alignment near Padamchen—Zuluk Sector, Project Swastik Project Swastik. Unpublished GSI report, FS 1973–1974

van Westen C (1997) Statistical landslide hazards analysis, ILWIS 2.1 for Windows application guide. ITC Publication, Enschede, pp 73–84

Chapter 10
Comparison Between Landslide Susceptibility Models: A Critical Review and Evaluation

10.1 Introduction

Among the various natural hazards, slope failure is the most widespread and damaging hazard (De Smedt 2005). A sudden failure of the slope is caused by sliding, rolling, falling or slumping. When failure occurs, material is transported down slope until a stable slope condition is re-established. The Darjeeling Himalayan terrain is very high susceptible to slope instability due to a complex geological structure and complex interaction among various processes acting upon the steep mountain southern escarpment slopes. In Darjeeling, the spatial extents of landslides are increasing day by day and causing severe damage to lives and properties. The Balason river basin is not an exception to it. The evolution of landforms is largely dependent on its geological structure, lithology, relief configuration, climatic characteristics and biophysical processes. The Balason river basin of Darjeeling Himalaya is an important geomorphic unit in regard to landslides. Researchers stated to find out sophisticated techniques and methods to study the landslides considering its destructive characters and propounded various models i.e. slope stability models (SINMAP model, Limit Equilibrium Model, Mohr-Coulomb failure criterion and safety factor (FS)), hydrological models, and statistical models. In this book, different statistical models have been studied and landslide susceptibility zonation map of the Balason river basin has also been made in GIS environment. To perform all the statistical models, first of all landslide causative factors were determined and corresponding data layers were made. One of the prime requisite to perform statistical models is a landslide inventory of the concerned study area. The most widely used landslide inventory techniques include (Montgomery and Dietrich 1994): (i) field investigation using a check list to identify landslide susceptibility sites; (ii) projection of future pattern of instability from the landslide inventories; (iii) multivariate analysis of factors; (iv) stability ranking based on criteria such as slope, lithology, landform; and (v) failure probability analysis based on slope stability models with stochastic hydrologic

S. Mandal and S. Mondal, *Geoinformatics and Modelling of Landslide Susceptibility and Risk*, Environmental Science and Engineering,
https://doi.org/10.1007/978-3-030-10495-5_10

simulations. Recently, the availability of GIS data has provided a lot of advantages to quantify topographic attributes related to slope instability and landsliding.

The growth of human population and expansion of human settlement at an alarming rate and its impact put enormous pressure in heavily weathered earth materials over mountain slope. This situation helps to move earth materials downward and cause slope failure. The Balason river basin faced so many landslide activities which damaged settlement area, communication lines and tea garden area. To check the destructive impact of landslide activities, researchers thought that landslide hazard zonation map may be an important techniques which will identify potential landslide locations for planning and development as well as the management of hill slope. This book highlighted different morphometric parameters, drainage diversity, relief diversity, morphometric diversity, tectonic and hydrologic attributes of the Balason river basin thoroughly and tried to find out the landslide potentiality in regard to each parameters. In the present book, 15 landslide causative factors such as elevation, aspect, slope, curvature, geology, soil, lineament density, distance to lineament, drainage density (DD), distance to drainage, stream power index (SPI), topographic wetted index (TWI), rainfall, normalized differential vegetation index (NDVI) and land use and land cover (LULC) has been taken into account and several statistical models have been performed to integrate all the data layers and to develop landslide susceptibility zonation maps. To develop landslide susceptibility zonation mal various models i.e. *information value* (IV) *method* after van Westen (1997), Chen et al. (2014); *landslide normal risk factor* (LNRF) *model* after Gupta and Joshi (1990), Torkashvand et al. (2014); *weights of Evidence* (WoE) *model* after Pradhan et al. (2010), Pourghasemi et al. (2013); *statistical index* (SI) *model* after Yalcin et al. (2011), Pourghasemi et al. (2013); *fuzzy Logic* (FL) after Bui et al. (2012), Sharma et al. (2013); *certainty factor* (CF) *model* after Sujatha et al. (2012), Pourghasemi et al. (2013); *support vector machine* (SVM) after Pradhan (2013) and Hong et al. (2015); and *artificial neural networks* (ANN) *model* after Pradhan and Lee (2010) and Choi et al. (2012) were being adopted. Every model has its own significance but it varies in terms of accuracy level. All these models expounded an individual outcome with regard to spatial distribution of landslide susceptibility. The models should be developed in such a way which can provide high level of accuracy and may be accepted for planning and development.

10.2 Landslides Susceptibility Studies: A Methodological Review

The present book comprises ten chapters out of which eight chapters are closely associated with various techniques and methods of landslide studies. Chapter 2 is dealt with the role of various geomorphic, geo-tectonic and hydrologic attributes i.e. elevation, slope, aspect, curvature, geomorphology, geology, lineament density, distance to lineament, land use/land cover (LULC), soil, normalized differential vegetation index (NDVI), drainage density, distance to drainage, stream power index (SPI), topographic wetted index (TWI) and rainfall in landslide occurrences in the Balason river basin of Darjeeling Himalaya. The probability of landslides of each class of the landslide causative factors were estimated with the help of *frequency ratio* (FR) which is derived by landslide affected pixels and landslide non-affected pixels. The experiences about the concerned study area revealed that the amount and intensity of rainfall, drainage network, lithological composition, soil character and slope character plays a significant role to cause landslide events in the Darjeeling Himalaya.

Chapter 3 focuses on sub-watersheds wise slope instability analysis and prioritization based on the 13 morphometric parameters of the Balason river basin, Darjeeling Himalaya, West Bengal, India, using RS and GIS techniques. Prioritization rating of 18 sub-watersheds is carried out through compound ranking method. The sub-watershed with the lowest rank is given the highest priority in terms of erosion and suggested for conservation measures urgently. Morphometric analysis of any drainage basin has paid immense importance in hydrological investigation. Morphometric parameters describe the topology, the structure, the planform and the relief of basin which are being applied for the prioritization of watersheds. The toposheets of 78A/4, 78A/8, 78B/1, 78B/5 and 78B/6 (Scale 1:50000), Google earth, Arc GIS 9.3 software and MS excel-2007 were used to assess 13 morphometric parameters i.e. bifurcation ratio (Rb), length of overland flow (Lof), drainage density (Dd), stream frequency (Fs), texture ratio (Rt), drainage texture (Td), compactness coefficient (Cc), constant of channel maintenance (Ccm), shape factor/basin shape (Sf), form factor (Ff), circularity ratio (Rc), elongation ratio (Re) and relief ratio (Rr) of the Balason River basin of Darjeeling Himalaya and to prioritize sub-watersheds (SW). To find out the sub-watersheds which are more unstable in regard to slope instability two major steps have been considered i.e. (i) *assessment of morphometric parameters* and *preparation of spatial data layers* and (ii) *slope instability analysis* and *sub-watersheds wise prioritization using compound ranking method*. The results showed that SW 4, SW 5 and SW 8 fall under very high priority class in respect of soil erosion and soil loss or simply instability having Rb of 3.532–4.002, Dd of 5.986–6.538, Fs of 14.934–29.447, Td of 9.765–23.158, Rt of 7.055–16.958, Lof of 0.069–0.084, Rr of 0.209–0.302, Cc of 1.265–1.554, Ccm of 0.139–0.167, Sb of 2.969–5.556, Ff of 0.180–0.337, Rc of 0.420–0.634 and Re of 0.479–0.655. Sub-watershed wise highest priority obtained by SW 5 followed by SW 8, SW 4, SW3, SW 9, SW 1, SW 6, SW 16, SW 2, SW 17, SW 15, SW 13, SW 7, SW 10, SW 11, SW 12, SW 14, and SW 18.

Chapter 4 described the role of basin morphometric parameters in slope instability using morphometric diversity (MD) model. Also try to find out the role of drainage parameters and relief parameters in slope failure using drainage diversity (DD) and relief diversity (RD) models respectively. For that total 14 morphometric data layers were considered. The relationship of each data layers with landslide susceptibility was judge using frequency ratio (FR) approach. Parameters like drainage density, drainage frequency, relative relief, drainage texture, junction frequency, infiltration number, ruggedness index, dissection index, elevation, slope, relief ratio and hypsometric integral were positively related with landslide potentiality while bifurcation ratio and drainage intensity negatively correlated with slope failure. The principal component analysis (PCA) based weight assign to each data layers of each model which multiplied with unidirectional reclassified data layers for each model using weighted linear combination (WLC) approach to prepare landslide susceptibility maps. A total of 295 landslide locations were which were randomly divided into two parts i.e. 70% of the total landslide locations were used as training data set and the remaining 30% locations were used for validation of the model.

Chapter 5 encompasses the application of Information value model (IVM), landslide nominal risk factor mode (LNRFM), fuzzy logic approach (FLA) and statistical index model (SIM) and the preparation of landslide susceptibility maps of the Balason river basin of Darjeeling Himalaya using various geomorphic, hydrologic, and tectonic attributes such as elevation, slope, aspect, curvature, geology, geomorphology, soil, distance to lineament, lineament density, drainage density, distance to drainage, stream power index (SPI), topographic wetness index (TWI), land use and land cover (LULC) and NDVU. All the landslide conditioning factors were being processed in GIS platform.

Chapter 6 discussed the implication of *weight of evidence model* and *certainty factor model* to prepare landslide susceptibility maps of the Balason river basin of Darjeeling Himalaya using data layers of elevation, slope, aspect, curvature, geology, geomorphology, soil, distance to lineament, lineament density, drainage density, distance to drainage, stream power index (SPI), topographic wetness index (TWI), land use and land cover (LULC) and NDVU in ARC GIS 10.1. The developed landslide susceptibility map was classified in five i.e. very low, low, moderate, high and very high landslide susceptibility. The prepared landslide susceptibility maps were also validated using ROC curve which stated that certainty factor mode is best suited for developing landslide susceptibility zonation map of the Balason river basin of Darjeeling Himalaya.

Chapter 7 is dealt with the preparation of landslide susceptibility map of the Balason river basin of Darjeeling Himalaya with the help of GIS tools machine learning model i.e. support vector machine (SVM) and artificial neural network model (ANNM). Fifteen landslide causative factors i.e. slope, aspect, curvature, elevation, geology, geomorphology, soil, distance to drainage, drainage density, distance to lineaments, lineament density, land use and land cover, stream power index (SPI), topographic wetness index (TWI) and rainfall were considered to produce the landslide susceptibility zonation maps SVM classification algorithm was performed for each factor by using the RBF kernel to prepare landslide susceptibility map. And, the back-propagation method was also applied to estimate factor's weight and the landslide hazard indices were derived with the help of trained back-propagation weights using ANN model. To validate the prepared landslide susceptibility map landslide inventory was used and accuracy result was obtained after processing ROC curve.

Chapter 8 attempted to assess geo-spatial distribution of landslide susceptibility in the Balason river basin of Darjeeling Himalaya using clustering of various factors i.e. geomorphological factors, lithological factors group, hydrological factors, triggering factor, protective factor and anthropogenic factor. The geomorphological factors, lithological factors group, hydrological factors, triggering factor, protective factor and anthropogenic factor were being integrated with the help of *geomorphological factor group model, lithological factor group model, hydrological factor group model, triggering factor group model, protective factor group model* and *anthropogenic factor group model*. To prepare data layers such as elevation, slope, aspect, curvature, geology, geomorphology, soil, drainage density, distance to drainage, lineament density, distance to lineament, stream power index, topographic wetness index, NDVI and LULC Google earth, topographical maps, SRTM DEM, satellite images were processed properly in GIS environment.

In Chap. 9, fifteen landslide inducing factors were considered i.e. elevation, aspect, slope, curvature, geology, soil, lineament density, distance to lineament, drainage density, distance to drainage, stream power index (SPI), topographic wetted index (TWI), rainfall (Ra), normalized differential vegetation index (NDVI) and land use and land cover (LULC) for the preparation landslide susceptibility zonation maps using CF model. The assessment of different susceptibility categories of settlement area and metalled road were done based on the aforesaid susceptibility zonation method. The elements at risk data layers i.e. settlement areas and roads were digitized from the maps which downloaded from www.banglarbhumi.gov.in website. The aforesaid model based susceptibility index map was intersected with the elements at risk data layers in Arc GIS 10.1 environment to obtain different susceptibility values of settlement area and road length.

10.3 Comparison Between Various Models Based on ROC Analysis

This book is accompanied with information value model, landslide nominal risk factor model, fuzzy logic approach, statistical index model, weight of evidence model, certainty factor model, support vector machine, artificial neural network model, and clustering model and the preparation of landslide susceptibility zonation map of the Balason river basin of Darjeeling Himalaya. Several researchers used ROC curve for the accuracy assessment of landslide susceptibility map (Bui et al. 2012; Ozdemir and Altural 2013). The ROC curve is the graphical presentation of the sensitivity on the y axis and 1-specificity of the model on the x axis (Mohammady et al. 2012). The area under curve (AUC) value of ROC curve varies from 0.5 to 1.0, values below 0.5 indicates a random fit and values close to 1 means excellent prediction accuracy.

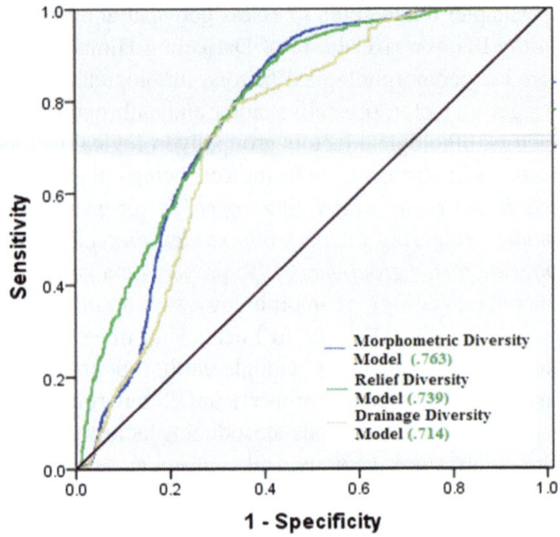

Fig. 10.1 ROC curve for drainage diversity model, relief diversity model and morphometric diversity model

10.3.1 Comparison Between Drainage Diversity Model, Relief Diversity Model and Morphometric Diversity Model

The area under curve (AUC) value of receiver operating characteristics (ROC) curve indicated that the prediction accuracy of the *drainage diversity model (DDM)*, *relief diversity model* (RDM) and *morphometric diversity model* (MDM) was 71.4, 73.9 and 76.3% respectively (Fig. 10.1). Three DD, RR and MD models yield reasonable prediction accuracy and can be used for identifying the role of drainage, relief and morphometric parameters in landslide susceptibility assessment and prediction.

10.3.2 Comparison Between Information Value Model, Landslide Nominal Risk Factor Model, Fuzzy Logic Approach and Statistical Index Model

The area under curve (AUC) value of the prepared LSI maps were 0.744 or 74% for information value mode (IVM), 0.742 or 74% for landslide nominal risk factor model (LNRFM), 0.975 or 97% for fuzzy logic approach (FLA) and 0.797 or 79% for statistical index model (SIM). The result of ROC curve depicts that fuzzy logic approach has high % accuracy (97%) and may be accepted as one of the most significant bivariate model in landslide susceptibility studies of the Balason river basin of Darjeeling Himalaya (Fig. 10.2).

Fig. 10.2 ROC curve for MIV model, LNRF model, FL and SI model

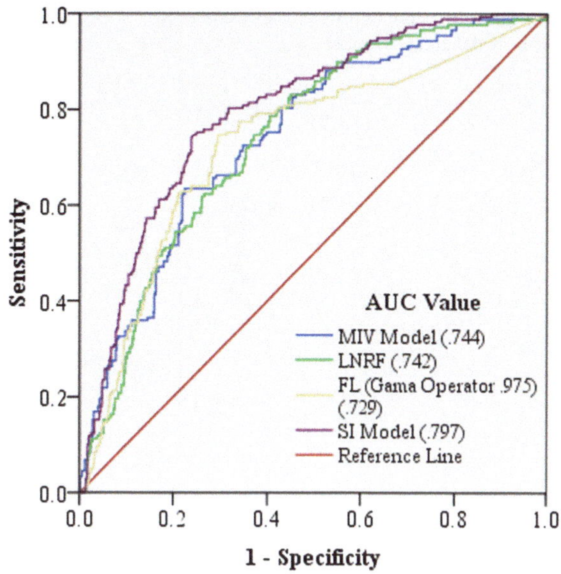

10.3.3 Comparison Between Weight of Evidence Model and Certainty Factor Model

The area under curve (AUC) value of the prepared LSI maps was 0.756 or 75% for weight of evidence model (WoEM) and 0.849 or 84% for certainty factor model (CFM). The result of ROC curve depicts that certainty factor model (CFM) has high % accuracy (84%) and may be accepted as one of the most significant bivariate model in landslide susceptibility studies of the Balason river basin of Darjeeling Himalaya (Fig. 10.3).

10.3.4 Comparison Between Support Vector Machine (SVM) and Artificial Neural Network (ANN) Model

An accuracy study has been made to assess the validation of landslide susceptibility maps. In this study ROC has been used for measuring the accuracy level of the landslide susceptibility zonation map. The ROC curve is the graphical presentation of the sensitivity on the y axis and 1-specificity of the model on the x axis (Mohammady et al. 2012). The area under curve for SVM based landslide susceptibility zonation map is 0.841 (84.1%) and for ANN model based landslide susceptibility map is 0.811 (81.1%) (Fig. 10.4). Between SVM and ANN model, SVM model depicted high level of accuracy and this model can be used for the preparation of landslide susceptibility zonation map.

Fig. 10.3 ROC curve for WoE model and CF model

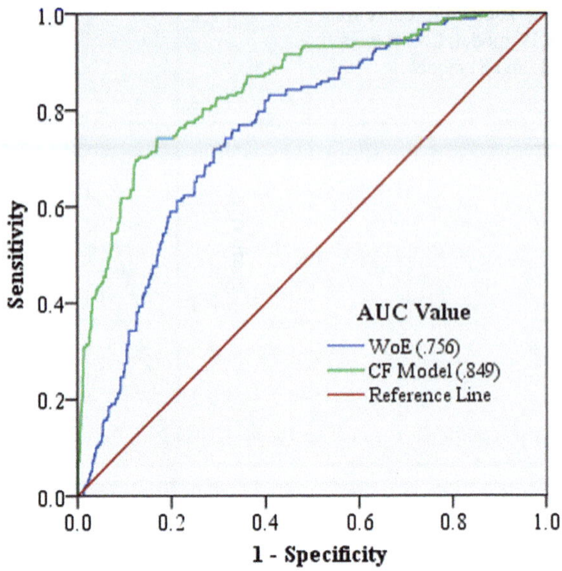

Fig. 10.4 ROC curve for SVM model and ANN model

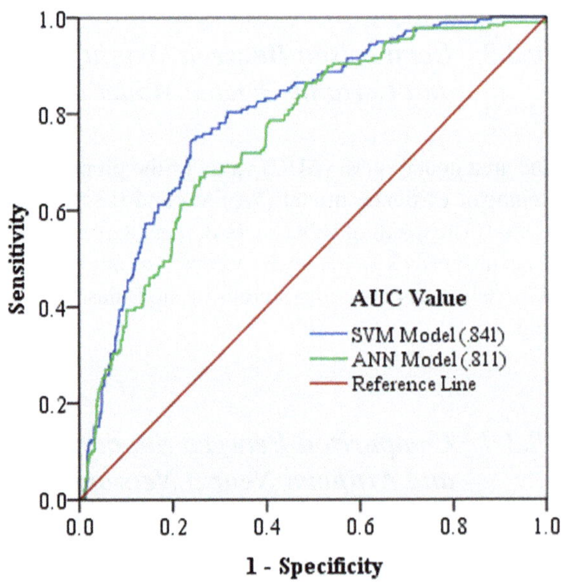

10.3.5 Comparison Between Factors Group Models

It can be concluded that each model can be treated as unique in terms of terrain processes and forms and can be used to identify role of geomorphological factors group model, lithological factors group model, hydrological factors group model, triggering factor model, protective factor model and anthropogenic factor model in slope failures individually. Lithological factors group model was played most prominent role to determine the landslides vulnerability zones as indicated by correlation value of 0.86 followed by geomorphological factors group model (0.85), hydrological factor group model (0.83), anthropogenic factor model (0.67), protective factor model (0.66) and triggering factor model (0.40).

10.4 Conclusion

The area under ROC curve of all the models i.e. information value model, landslide nominal risk factor model, fuzzy logic approach, statistical index model, weight of evidence model, certainty factor model, support vector machine, and artificial neural network model are 74.4, 74.2, 72.9 (fuzzy gamma operator 97.5), 79.7, 75.6, 84.9, 84.1 and 81.1% respectively. The ROC analysis revealed a high level of accuracy of certainty factor model (84.9%) which rank first in terms of accuracy and followed by support vector machine, ANN model, statistical index model, weight of evidence model etc. Pradhan (2013) studied landslide susceptibility using *support vector machine* and found the accuracy level of more than 80% which showed a close coincidence of the accuracy level of 84.1% as found for support vector machine in the present study. In ANN model, the AUC value for the model with six factors (0.82) is slightly higher than that with fourteen factors (0.73) as expounded by Dou et al. (2015) which differ from AUC value of 81.1% adopted in the study. Devkota et al. (2013) applied certainty factor model (CFM) to prepare landslide susceptibility zonation map and plotted ROC which performed accuracy level of 87.21%. The accuracy level of 84.4% in CFM of the study is very close to the accuracy level of 87.21% after Devkota et al. (2013). The area under ROC curve estimated using WoE model by Chen and Li (2014) is 72.00% and area under ROC curve of the study is 75.6%. The study explored that certainty factor model is one of the most significant statistical model in landslide susceptibility assessment and prediction because of its high % of accuracy not in the present study but also in other landslide susceptibility studies. In terms of factor's group model, *lithological factors group model* (0.86) ranks first and followed by geomorphological factors group model (0.85), hydrological factor group model (0.83), anthropogenic factor model (0.67), protective factor model (0.66) and triggering factor model (0.40). It can be summarized that preparation of landslide causative factors data layers and landslide inventory win full care and identification suitable models to assess landslide susceptibility which will help a lot to the planners and developers to develop settlement area, tea garden area, communication lines etc. in the mountain terrain of Darjeeling Himalaya where lies the significance of landslide susceptibility zonation map and statistical models.

References

Bui DT, Pradhan B, Lofman O, Revhaug I, Dick OB (2012) Spatial prediction of landslide hazards in Hoa Binh province (Vietnam): a comparative assessment of the efficacy of evidential belief functions and fuzzy logic models. Catena 96:28–40

Chen W, Li W (2014) Application of weights-of-evidence model in landslide susceptibility mapping at Baozhong Region in Baoji, China. Electron J Geotech Eng 19

Chen W, Li W, Hou E (2014) Landslide susceptibility mapping based on GIS and information value model for the Chencang District of Baoji, China. Arab J Geosci 7:4499–4511

Choi J, Oh H-J, Lee H-J, Lee C, Lee S (2012) Combining landslide susceptibility maps obtained from frequency ratio, logistic regression, and artificial neural network models using ASTER images and GIS. Eng Geol 124:12–23. https://doi.org/10.1016/j.enggeo.2011.09.011

Devkota KC, Regmi AD, Yoshida KH, Pourghasemi R, Pradhan B, Althuwaynee OF, Dhital MR (2013) Landslide susceptibility mapping using certainty factor, index of entropy and logistic regression models in GIS and their comparison at Mugling-Narayanghat road section in Nepal Himalaya. Nat Hazards 65:135–165. https://doi.org/10.1007/s11069-012-0347-6

De Smedt F (2005) Slope stability analysis using GIS on a regional scale: a case study of Narayanghat Mungling highway section, Nepal. Dissertation of Master Degree, Universiteit Gent, Belgium

Dou J, Yamagishi H, Pourghasemi HR, Song X, Xu Y, Zhu Z (2015) An integrated artificial neural network model for the landslide susceptibility assessment of Osado Island, Japan. Nat Hazards 78:1749–1776. https://doi.org/10.1007/s11069-015-1799-2

Gupta RP, Joshi BC (1990) Landslide hazard zonation using the GIS approach—a case study from the Ramganga Catchment, Himalayas. Eng Geol 28:119–131

Hong H, Pradhan B, Xu C, Bui DT (2015) Spatial prediction of landslide hazard at the Yihuang area (China) using two-class kernel logistic regression, alternating decision tree and support vector machines. Catena 133:266–281

Mohammady M, Pourghasemi HR, Pradhan B (2012) Landslide susceptibility mapping at Golestan Province, Iran: a comparison between frequency ratio, Dempster-Shafer, and weights-of-evidence models. J Asian Earth Sci 61:221–236

Montgomery DR, Dietrich WE (1994) A physically based model for the topographic control on shallow land sliding. Water Resour Res 30(4):1153–1171

Ozdemir A, Altural T (2013) A comparative study of frequency ratio, weights of evidence and logistic regression methods for landslide susceptibility mapping: Sultan Mountains, SW Turkey. J Asian Earth Sci 64:180–197

Pradhan B (2013) A comparative study on the predictive ability of the decision tree, support vector machine and neuro-fuzzy models in landslide susceptibility mapping using GIS. Comput Geosci 51:350–365

Pradhan B, Lee S (2010) Landslide susceptibility assessment and factor effect analysis: backpropagation artificial neural networks and their comparison with frequency ratio and bivariate logistic regression modelling. Environ Model Softw 25(6):747–759

Pradhan B, Oh HJ, Buchroithner M (2010) Weights-of-evidence model applied to landslide susceptibility mapping in a tropical hilly area. Geomat Nat Hazards Risk 1(3):199–223

Pourghasemi HR, Pradhan B, Gokceoglu C, Mohammadi M, Moradi HR (2013) Application of weights-of-evidence and certainty factor models and their comparison in landslide susceptibility mapping at Haraz watershed, Iran. Arab J Geosci 6(7):2351–2365

Sharma LP, Patel N, Ghose MK, Debnath P (2013) Synergistic application of fuzzy logic and geoinformatics for landslide vulnerability zonation—a case study in Sikkim Himalayas, India. Appl Geomat 5:271–284

Sujatha ER, Rajamanickam GV, Kumaravel P (2012) Landslide susceptibility analysis using probabilistic certainty factor approach: a case study on Tevankarai stream watershed, India. J Earth Syst Sci 121(5):1337–1350

Torkashvand AM, Irani A, Sorur J (2014) The preparation of landslide map by Landslide Numerical Risk Factor (LNRF) model and Geographic Information System (GIS). Egypt J Remote Sens Space Sci 17:159–170

van Westen CJ (1997) Statistical landslide hazard analysis. In: Application guide, ILWIS 2.1 for Windows. ITC, Enschede, The Netherlands, pp 73–84

Yalcin A, Reis S, Aydinoglu AC, Yomralioglu T (2011) A GIS-based comparative study of frequency ratio, analytical hierarchy process, bivariate statistics and logistics regression methods for landslide susceptibility mapping in Trabzon, NE Turkey. Catena 85(3):274–287

FSC
www.fsc.org
MIX
Papier aus verantwortungsvollen Quellen
Paper from responsible sources
FSC® C105338

If you have any concerns about our products,
you can contact us on
ProductSafety@springernature.com

In case Publisher is established outside the EU,
the EU authorized representative is:
Springer Nature Customer Service Center GmbH
Europaplatz 3, 69115 Heidelberg, Germany

Printed by Libri Plureos GmbH
in Hamburg, Germany